CLASSIC SERIES

Greatest.
Science Fiction
Stories

V&S PUBLISHERS

Published by:

V&S PUBLISHERS

F-2/16, Ansari road, Daryaganj, New Delhi-110002
☎ 23240026, 23240027 • *Fax:* 011-23240028
Email: info@vspublishers.com • *Website:* www.vspublishers.com

Regional Office : Hyderabad
5-1-707/1, Brij Bhawan (Beside Central Bank of India Lane)
Bank Street, Koti, Hyderabad - 500 095
☎ 040-24737290
E-mail: vspublishershyd@gmail.com

Branch Office : Mumbai
Flat No. Ground Floor, Sonmegh Building
No. 51, Karel Wadi, Thakurdwar, Mumbai - 400 002
☎ 022-22098268
E-mail: vspublishersmum@gmail.com

Follow us on: 🇹 f in

For any assistance sms **VSPUB** to **56161**

All books available at **www.vspublishers.com**

© **Copyright:** V&S PUBLISHERS
ISBN 978-93-505710-5-7
Edition 2014

Printed at : Param Offseters, Okhla, New Delhi-110020

Publisher's Note

It has been our constant endeavour at the **V&S Publishers** to publish all kinds of books ranging from Fiction, Non-fiction, Storybooks, Children Encyclopaedias, to Self-Help, Science Books, Dictionaries, Grammar Books, Self-Development, Management Books, etc.

However, this is for the first time that we are venturing into the vast, rich and fathomless ocean of English Literature and have come up with a set *of ten storybooks called the Greatest Classic Series* authored by some of the greatest and eminent writers of the world. There is a lot to learn from their writing style, selection of plot, development and building of theme and suspense of the story, emphasis and presentation of characters, dialogues, working towards the climax of the story, presenting the climax, and then finally concluding the story.

Each these books are of about 200 pages containing around ten popular stories or more of renowned authors like Oscar Wilde, Ernest William Hornung, Guy de Maupassant, O. Henry, Saki, Washington Irving, Thomas Hardy, Charles Dickens, Jules Verne, Jack London, Mark Twain, Edgar Allen Poe, H.G.Wells, Ambrose Bierce, Amelia Edwards, Edith Wharton, Wilkie Collins and many more. The series is called The Greatest Classic Series as all the names of the books begin with the word, `Greatest' like the Greatest Adventurous Stories, Greatest Detective Stories, Greatest Love Stories, Greatest Ghost Stories, and so on. Besides this, three of the ten books are exclusively on the Adventures of Sherlock Holmes, one of the best detectives the world has ever known written by none other than Sir Arthur Conan Doyle.

Besides the above mentioned characteristics, the books contain an introductory page before each story introducing the author, his brief life history, notable works and literary achievements. Each story has a set of word meanings on each page followed by an exercise meant exclusively aiming the school students to help them grasp the essence of the story easily and quickly.

These books are not only a boon for the school-going students, particularly studying in senior classes from the seventh standard till the twelfth, but are also a treasure trove for all those young and aspiring writers, voracious readers and lovers of English language and literature.

Each of these ten books focus on a theme, such as adventure, love, terror, humour, or supernatural happenings, and are so captivating and real to life that readers may find it difficult to choose from them and so it's better to pick the entire series.

Wishing you all a happy and enjoyable reading...

Contents

Publisher's Note 3

Jules Verne 7

The Day of an American Journalist in 2889 9

Tobermory 29

'Ministers of Grace' 39

The Vampyre 53

Manuscript Found in a Bottle 76

The Enemy of All the World 91

The Mummy's Foot 110

The Terror of Blue John Gap 127

The Haunted House 147

One Of Twins 180

Moxon's Master 189

Jules Verne

Born on February 8, 1828

Died on March 24, 1905

Notable Works: *Twenty Thousand Leagues Under the Sea, A Journey to the Center of the Earth, Around the World in Eighty Days, The Mysterious Island, Dick Sand, A Captain at Fifteen, etc.*

Genres: Science fiction

Early Life

Jules Gabriel Verne was a French author, who pioneered the **science fiction genre**. He is best known for his novels, *Twenty Thousand Leagues Under the Sea* (1870), *A Journey to the Center of the Earth* (1864), and *Around the World in Eighty Days* (1873). Verne wrote about space, air, and underwater travel before air travel and practical submarines were invented, and before practical means of space travel had been devised. He is the second most translated author in the world (after Agatha Christie). Some of his books have also been made into live-action and animated films and television shows. Verne is often referred to as the "Father of Science Fiction", a title sometimes shared with Hugo Gernsback and H. G. Wells.

Jules Verne was born in Nantes, in France, to Pierre Verne, an attorney, and his wife, Sophie Allote de la Fuÿe. Jules spent his early years at home with his parents in the bustling harbour city of Nantes. The family spent the summers in a country house, just outside the city, in Brains on the banks of the Loire River. At the boarding school, Verne studied Latin, which he used in his short story, "Le Mariage de Monsieur Anselme des Tilleuls" in the mid 1850s.

Literary Works and Achievements

After completing his studies at the lycée, Verne went to Paris to study law. Around 1848, in conjunction with Michel Carré, he began writing libretti for operettas, five of them for his friend the composer, Aristide Hignard, who also set Verne's poems as chansons. For some years, his attentions were divided between the theatre and work, but some travellers' stories which he wrote for the Musée des familles revealed to him his true talent. When Verne's father discovered that his son was writing rather than studying law, he promptly withdrew his financial support.

It was Jules Verne himself who confirmed once having travelled from Stockholm, Sweden, to Christiania during 1862. Consequently, Swedish publication of Jules Verne began during the very next year. *En luftballongsresa genom Afrika* (A Hot Air Balloon trip through Africa), translated into Swedish, actually is dated 1863 – making it the very first dated Jules Verne full-length book on record.

Verne's situation improved when he met Pierre-Jules Hetzel, one of the more important French publishers of the 19th century. From that point, Verne published two or more volumes a year. The most successful of these include *Voyage au centre de la Terre* (Journey to the Center of the Earth, (1864); *De la Terre à la Lune* (From the Earth to the Moon, 1865); *Vingt mille lieues sous les mers* (Twenty Thousand Leagues Under the Sea, 1869); and *Le tour du monde en quatre-vingts jours* (Around the World in Eighty Days), which first appeared in *Le Temps* in 1872. The series is collectively known as "Voyages Extraordinaires" ("extraordinary voyages").

Verne could now live on his writings. But most of his wealth came from the stage adaptations of *Le tour du monde en quatre-vingts jours* (1874) and *Michel Strogoff* (1876), which he wrote with Adolphe d'Ennery. In 1867, Verne bought a small ship, the Saint-Michel, which he successively replaced with the Saint-Michel II and the Saint-Michel III as his financial situation improved. In 1863, Jules Verne wrote a novel called *Paris in the Twentieth Century.*

Later Years

On March 9, 1886, as Verne was coming home, his twenty-five-year-old nephew, Gaston, shot him with a gun. One bullet missed, but the second entered Verne's left leg, giving him a limp that would not be cured. The incident was hushed up by the media, and Gaston spent the rest of his life in an asylum.

After the deaths of Hetzel and his beloved mother Sophie Henriette Allotte de la Fruye in 1887, Jules began writing darker works. In 1888, Jules Verne entered politics and was elected the town councillor of Amiens, where he championed several improvements and served for 15 years. In 1905, while ill with diabetes, Verne died at his home.

Trivia

A restaurant built into the Eiffel Tower in Paris, France is named *Le Jules Verne.*

The Day of an American Journalist in 2889

~ Jules Verne

THe men of the twenty-ninth century live in a *perpetual* fairyland, though they do not seem to realise it. Bored with wonders, they are cold towards everything that progress brings them every day. It all seems only natural.

If they compared it with the past, they would better appreciate what our civilisation is, and realise what a road it has **traversed**. What would then seem finer than our modern cities, with streets a hundred yards wide, with buildings a thousand feet high, always at an equable temperature, and the sky furrowed by thousands of aero-cars and aero-buses! Compared with these towns, whose population may include up to ten million *inhabitants*, what were those villages, those hamlets of a thousand years ago, that Paris, that London, that New York - muddy and badly ventilated townships, traversed by jolting contraptions, hauled along by horses - yes! by horses! it's unbelievable!

If they recalled the erratic working of the steamers and the railways, their many *collisions*, and their slowness, how greatly would travellers value the aero-trains, and especially these pneumatic tubes that laid beneath the oceans, which convey them with a speed of a thousand miles an hour? And would they not enjoy the telephone and the telephote even better if they recollected that our fathers were reduced to that antediluvial apparatus which they called the 'telegraph'? It's very strange. These surprising transformations are based on principles which were quite well-known to our ancestors, although these, so to speak, made no use of them. Heat, steam, electricity are as old as mankind. Towards the end of the nineteenth century, did not the savants declare that the only difference between the physical and chemical forces consists of the special rates of vibration of the etheric particles?

As so enormous a stride had been made, that of recognising the mutual relationship of all these forces, it is incredible that it took so long to work out the rates of vibration than differentiate between them. It is especially surprising that

Perpetual
– Continuous
Traverse *– Cross*
Inhabitant
– Occupant
Collision *– Crash*

the method of passing directly from one to another, and of producing one without the other, has only been discovered so recently. So it was, however, that things happened, and it was only in 2790, about a hundred years ago, that the famous Oswald Nyer succeeded in doing so.

A real *benefactor* of humanity, that great man! His achievement, a work of genius, was the parent of all the others! A constellation of inventors was born out of it, *culminating* in our extraordinary James Jackson. It is to him that we owe the new accumulators, some of which condense the force of the solar rays, others the electricity stored in the heart of our globe, and yet again others, energy coming from any source whatever, whether it be the waterfalls, winds, or rivers. It is to him that we owe no less the transformer which, at a touch on a simple switch, draws on the force that lives in the accumulators and releases it as heat, light, electricity, or mechanical power after it has performed any task we need.

Yes, it was from the day on which these two appliances were thought out that progress really dates. They have given mankind almost an infinite power. Through mitigating the *bleakness* of winter by restoring to it the excessive heat of the summer, they have revolutionised agriculture. By providing motive power for the appliances used in aerial *navigation*, they have enabled commerce to make a splendid leap forward. It is to them that we owe the unceasing production of electricity without either batteries or machines, light without combustion or incandescence, and finally that inexhaustible source of energy which has increased industrial production a hundredfold.

Very well then! The whole of these wonders, we shall meet them in an incomparable office block - the office of the Earth Herald, recently inaugurated in the 16823rd Avenue.

If the founder of the New York Herald, Gordon Bennett, were to be born a second time today, what would he say when he saw this palace of marble and gold that belongs to his *illustrious* descendant, Francis Bennett? Thirty generations had followed one another, and the New York Herald had always stayed in that same Bennett family. Two hundred years before, when the government of the Union had been transferred from Washington to Centropolis, the newspaper had followed the government – if it were not that the government had followed the newspaper – and it had taken its new title, the Earth Herald.

Benefactor – *Patron*
Culminate – *End*
Bleak – *Drab*
Navigation – *Direction-finding*
Illustrious – *Memorable*

And let nobody imagine that it had declined under the administration of Francis Bennett. No! On the contrary, its new director had given it an equalled vitality and driving-power by the inauguration of telephonic journalism.

Everybody knows that system, made possible by the incredible diffusion of the telephone. Every morning, instead of being printed as in antiquity, the Earth Herald is 'spoken'. It is by means of a **brisk** conversation with a reporter, a political figure, or a scientist, that the subscribers can learn whatever happens to interest them. As for those who buy an odd number for a few cents, they know that they can get acquainted with the day's issue through the countless phonographic cabinets.

This innovation of Francis Bennett restored new life to the old journal. In a few months its clientele numbered eighty-five million subscribers, and the director's **fortune** rose to three hundred million dollars, and has since gone far beyond that. Thanks to this fortune, he was able to build his new office — a colossal **edifice** with four facades each two miles long, whose roof is sheltered beneath the glorious flag, with its seventy-five stars, of the Confederation.

Francis Bennett, king of journalists, would then have been king of the two Americas, if the Americans would ever accept any monarch whatsoever. Do you doubt this? But the plenipotentiaries of every nation and our very ministers throng around his door, peddling their advice, seeking his approval, imploring the support of his all-powerful organ. Count up the scientists whom he has encouraged, the artists whom he employs, the inventors whom he subsidises! A wearisome monarchy was his, work without **respite**, and certainly nobody of earlier times would ever have been able to carry out so unremitting a daily grind. Fortunately, however, the men of today have a more robust constitution, thanks to the progress of hygiene and of gymnastics, which from thirty-seven years has now increased to sixty-eight the average length of human life – thanks too to the **aseptic** foods, while we wait for the next discovery: that of nutritious air which will enable us to take nourishment. . .only by breathing.

And now, if you would like to know everything that constitutes the day of a director of the Earth Herald, take the trouble to follow him in his multifarious operations – this very day, this July 25th of the present year, 2889.

Brisk – *Hurried*
Fortune – *Wealth*
Edifice –
Organization
Respite – *Relief*
Aseptic – *Sterile*

That morning Francis Bennett awoke in rather a bad temper. This was eight days since his wife had been in France and he was feeling a little lonely. Can it be credited? They had been married ten years, and this was the first time that Mrs. Edith Bennett, that professional beauty, had been so long away. Two or three days usually **sufficed** for her frequent journeys to Europe and especially to Paris, where she went to buy her hats.

As soon as he awoke, Francis Bennett switched on his phonotelephote, whose wires led to the house he owned in the Champs-Elystes.

The telephone, completed by the telephote, is another of our time's **conquests**! Though the transmission of speech by the electric current was already very old, it was only since yesterday that vision could also be transmitted. A valuable discovery, and Francis Bennett was by no means the only one to bless its inventor when, in spite of the enormos distance between them, he saw his wife appear in the telephotic mirror.

A lovely vision! A little tired by last night's theatre or dance, Mrs. Bennett was still in bed. Although where she was it was nearly noon, her charming head was buried in the lace of the pillow. But there she was stirring . . . her lips were moving . . . No doubt she was dreaming? . . . Yes! She was dreaming . . . A name slipped from her mouth. 'Francis . . . dear Francis! . . .'

His name, spoken by that sweet voice, gave a happier turn to Francis Bennett's mood. Not wanting to wake the pretty sleeper, he quickly jumped out of bed, and went into his mechanised dressing room.

Two minutes later, without needing the help of a valet, the machine deposited him, washed, shaved, shod, dressed, and buttoned from top to toe, on the **threshold** of his office. The day's work was going to begin.

It was into the room of the serialised novelists that Francis first entered.

Very big, that room, **surmounted** by a large translucent dome. In a corner, several telephonic instruments by which the hundred authors of the Earth Herald related a hundred chapters of a hundred romances to the enfevered public.

Catching sight of one of these serialists who was snatching five minutes' rest, Francis Bennett said: 'Very fine, my dear

Suffice – *Serve*
Conquest – *Downfall*
Threshold – *Edge*
Surmount – *Defeat*

fellow, very fine, that last chapter of yours! That scene where the young village girl is discussing with her admirer some of the problems of transcendental philosophy shows very keen powers of observation! These country manners have never been more clearly depicted! Go on that way, my dear Archibald, and good luck to you. Ten thousand new subscribers since yesterday, thanks to you!

'Mr. John Last,' he continued, turning towards another of his collaborators, 'I'm not so satisfied with you! It hasn't any life, your story! You're in too much of a hurry to get to the end! Well! and what about all that documentation? You've got to dissect, John Last, you've got to dissect! It isn't with a pen one writes nowadays, it's with a scalpel! Every action in real life is the resultant of a succession of fleeting thoughts, and they've got to be carefully set out to create a living being! And what's easier than to use electrical hypnotism, which redoubles its subject and separates his twofold personality! Watch yourself living, John Last, my dear fellow! Imitate your colleague whom I've just been congratulating! Get yourself hypnotised...What?...You're having it done, you say?...Not good enough yet, not good enough!'

Having given this little lesson, Francis Bennett continued his inspection and went on into the reporters' room. His fifteen hundred reporters, placed before an equal number of telephones, were passing on to subscribers the news which had come in during the night from the four quarters of the earth.

The organisation of this incomparable service has often been described. In addition to his telephone, each reporter has in front of him a series of commutators, which allow him to get into communication with this or that telephotic line. Thus the subscribers have not only the story but the sight of these events. When it is a question of miscellaneous facts', which are things of the past by the time they are described, their principal phases alone are transmitted; these are obtained by intensive photography.

Depict – *Portray*
Imitate – *Emulate*
Inspection – *Review*
Miscellaneous –
Various
Intensive –
Concentrated

Francis Bennett questioned one of the ten astronomical reporters a service which was growing because of the recent discoveries in the stellar world.

'Well, Cash, what have you got?'

'Phototelegrams from Mercury, Venus, and Mars, sir.'

'Interesting, that last one?'

'Yes! a revolution in the Central Empire, in support of the reactionary liberals against the republican conservatives.'

'Just like us, then! And Jupiter?'

'Nothing so far! We haven't been able to understand the signals the Jovians make. Perhaps ours haven't reached them? . . .'

'That's your job, and I hold you responsible, Mr. Cash!' Francis Bennett replied; extremely dissatisfied, he went on to the scientific editorial room.

Bent over their computers, thirty savants were absorbed in equations of the ninety-fifth degree. Some indeed were revelling in the formulae of algebraical infinity and of twenty-four-dimensional space, like a child in the elementary class dealing with the four rules of arithmetic.

Francis Bennett fell among them rather like a bombshell.

'Well, gentlemen, what's this they tell me? No reply from Jupiter? . . . It's always the same! Look here, Corley, it seems to me it's been twenty years that you've been pegging away at that planet . . .'

'What do you expect, sir?' the savant replied. 'Our optical science still leaves something to be desired, and even with our telescopes two miles long . . .'

'You hear that, Peer?' broke in Francis Bennett, addressing himself to Corley's neighbour. 'Optical science leaves something to be desired! . . . That's your speciality, that is, my dear fellow! Put on your glasses, devil take it! Put on your glasses!'

Then, turning back to Corley:

But, failing Jupiter, aren't you getting some result from the moon, at any rate?

'Not yet, Mr. Bennett.'

'Well, this time, you can't blame optical science! The moon is six hundred times nearer than Mars, and yet our correspondence service is in regular operation with Mars. It can't be telescopes we're needing . . .'

'No, it's the inhabitants,' Corley replied with the thin smile of a savant stuffed with X.

'You dare tell me that the moon is uninhabited?'

'On the face it turns towards us, at any rate, Mr. Bennett. Who knows whether on the other side . . .?'

Savant – *Guru*
Revel – *Celebration*
Correspondence – *Communication*

'Well, there's a very simple method of finding out . . .'

'And that is? . . .'

'To turn the moon round!'

And that very day, the scientists of the Bennett factory started working out some mechanical means of turning our satellite right round.

On the whole Francis Bennett had reason to be satisfied. One of the Earth Herald's astronomers had just determined the elements of the new planet Gandini. It is at a distance of 12,841,348,284,623 metres and 7 decimetres that this planet describes its orbit round the sun in 572 years, 194 days, 12 hours, 43 minutes, 9.8 seconds. Francis Bennett was delighted with such precision.

'Good!' he exclaimed. 'Hurry up and tell the reportage service about it. You know what a passion the public has for these astronomical questions. I'm anxious for the news to appear in today's issue!'

Before leaving the reporters' room he took up another matter with a special group of interviewers, addressing the one who dealt with celebrities: 'You've interviewed President Wilcox?' he asked.

'Yes, Mr. Bennett, and I'm publishing the information that he's certainly suffering from a dilation of the stomach, and that he's most conscientiously undergoing a course of tubular irrigations.'

'Splendid. And that business of Chapmann the assassin . . . Have you interviewed the jurymen who are to sit at the Assizes?'

'Yes, and they all agree that he's guilty, so that the case won't even have to be submitted to them. The accused will be executed before he's sentenced.'

'Splendid! Splendid!'

The next room, a broad gallery about a quarter of a mile long, was devoted to publicity, and it well may be imagined what the publicity for such a journal as the Earth Herald had to be. It brought in a daily average of three million dollars. Very ingeniously, indeed, some of the publicity obtained took an absolutely novel form, the result of a patent bought at an outlay of three dollars from a poor devil who had since died of hunger. They are gigantic signs reflected on the clouds, so large that they can be seen all over a whole country. From that

Delight – *Pleasure*
Precise – *Exact*
Accuse – *Blame*
Gigantic – *Huge*

gallery a thousand projectors were unceasingly employed in sending to the clouds, on which they were reproduced in colour, these inordinate advertisements.

But that day when Francis Bennett entered the publicity room he found the technicians with their arms folded beside their idle projectors. He asked them about it . . . The only reply he got was that somebody pointed to the blue sky.

'Yes! . . . A fine day,' he muttered, 'so we can't get any aerial publicity! What's to be done about that? If there isn't any rain, we can produce it! But it isn't rain, it's clouds that we need!'

'Yes, some fine snow-white clouds!' replied the chief technician.

'Well, Mr. Simon Mark, you'd better get in touch with the scientific editors, meteorological service. You can tell them from me that they can get busy on the problem of artificial clouds. We really can't be at the mercy of the fine weather.'

After finishing his inspection of the different sections of the paper, Francis Bennett went to his reception hall, where he found awaiting him the ambassadors and plenipotentiary ministers accredited to the American government: these gentlemen had come to ask advice from the all-powerful director. As he entered the room they were carrying on rather a lively discussion. Pardon me, Your Excellency,' the French Ambassador addressed the Ambassador from Russia. 'But I can't see anything that needs changing in the map of Europe. The north to the Slavs, agreed! But the south to the Latins! Our common frontier along the Rhine seems quite satisfactory. Understand me clearly, that our government will certainly resist any attempt which may be made against our Prefectures of Rome, Madrid and Vienna!'

'Well said!' Francis Bennett intervened in the discussion. 'What, Mr. Russian Ambassador, you're not satisfied with your great empire, which extends from the banks of the Rhine as far as the frontiers of China? An empire whose immense coast is bathed by the Arctic Ocean, the Atlantic, the Black Sea, the Bosphorus, and the Indian Ocean!

'And besides, what's the use of threats? Is war with our modern weapons possible! These asphyxiating shells which can be sent a distance of a hundred miles, these electric flashes, sixty miles long, which can annihilate a whole army corps at

Unceasing – *Constant*
Inordinate – *Excessive*
Mercy – *Compassion*
Accredit – *Recognize*
Resist – *Fight*

a single blow, these projectiles loaded with the microbes of plague, cholera, and yellow fever, and which can destroy a whole nation in a few hours?'

'We realise that, Mr. Bennett,' the Russian Ambassador replied. 'But are we free to do what we like? . . . Thrust back ourselves by the Chinese on our eastern frontier, we must, at all costs, attempt something towards the west . . .'

'Is that all it is, sir?' Francis Bennett replied in reassuring tones - 'Well! as the proliferation of the Chinese is getting to be a danger to the world, we'll bring pressure to bear on the Son of Heaven. He'll simply have to impose a maximum birth-rate on his subjects, not to be exceeded on pain of death! A child too many? . . . A father less! That will keep things balanced.'

'And you, sir,' the director of the Earth Herald continued, addressing the English consul, 'what can I do to be of service to you?'

'A great deal, Mr. Bennett,' that personage replied. 'It would be enough for your journal to open a campaign on our behalf . . .'

'And with what purpose?'

'Merely to protest against the annexation of Great Britain by the United States . . .'

'Merely that!' Francis Bennett exclaimed. He shrugged his shoulders. 'An annexation that's a hundred and fifty years old already! But won't you English gentry ever resign yourselves to the fact that by a just compensation of events here below, your country has become an American colony? That's pure madness! How could your government ever have believed that I should even open so anti-patriotic a campaign?'

'Mr. Bennett, you know that the Monroe Doctrine is all America for the Americans, and nothing more than America, and not . . .'

'But England is only one of our colonies, one of the finest. Don't count upon our ever consenting to give her up.'

'You refuse? . . .'

'I refuse, and if you insist, we shall make it a casus belli, based on nothing more than an interview with one of our reporters.'

'So that's the end.' The consul was overwhelmed. 'The United Kingdom, Canada, and New Britain belong to the Americans, India to the Russians, and Australia and New

Thrust – *Shove*
Reassure – *Assure*
Proliferate – *Thrive*
Gentry – *Nobility*
Resign – *Leave*

Zealand to themselves! Of all that once was England, what's left? . . . Nothing'

'Nothing, sir?' retorted Francis Bennett. 'Well, what about Gibraltar?'

At that moment the clock struck twelve. The director of the Earth Herald, ending the audience with a gesture, left the hall, and sat down in a rolling armchair. In a few minutes he had reached his dining room half a mile away, at the far end of the office.

The table was laid, and he took his place at it. Within reach of his hand was placed a series of taps, and before him was the curved surface of a phonotelephote, on which appeared the dining room of his home in Paris. Mr. and Mrs. Bennett had arranged to have lunch at the same time - nothing could be more pleasant than to be face-to-face in spite of the distance, to see one another and talk by means of the phonotelephotic apparatus.

But the room in Paris was still empty.

'Edith is late,' Francis Bennett said to himself. 'Oh, women's punctuality! Everything makes progress, except that.'

And after this too just reflection, he turned on one of the taps.

Like everybody else in easy circumstances nowadays, Francis Bennett, having abandoned domestic cooking, is one of the subscribers to the Society for Supplying Food to the Home, which distributes dishes of a thousand types through a network of pneumatic tubes. This system is expensive, no doubt, but the cooking is better, and it has the advantage that it has suppressed that hair-raising race, the cooks of both sexes.

So, not without some regret, Francis Bennett was lunching in solitude. He was finishing his coffee when Mrs. Bennett, having got back home, appeared in the telephote screen.

'Where have you been, Edith dear?' Francis Bennett enquired.

'What' Mrs. Bennett replied. 'You've finished! . . . I must be late, then? . . . Where have I been? Of course, I've been with my modiste . . . This year's hats are so bewitching! They're not hats at all . . . they're domes, they're cupolas! I rather lost count of time'

'Rather, my dear? You lost it so much that here's my lunch finished.'

Retort – *Reply*
Gesture – *Signal*
Punctual – *On time*
Abandon – *Recklessness*
Solitude – *Loneliness*

'Well, run along then, my dear . . . run along to your work,' Mrs. Bennett replied. 'I've still got a visit to make, to my modeleur-couturier.'

And this couturier was no other than the famous Wormspire, the very man who so judiciously remarked, 'Woman is only a question of shape!'

Francis Bennett kissed Mrs. Bennett's cheek on the tele-phote screen and went across to the window, where his aero-car was waiting.

'Where are we going, sir?' asked the aero-coachman. 'Let's see. I've got time . . .' Francis Bennett replied. 'Take me to my accumulator works at Niagara.'

The aero-car, an apparatus splendidly based on the prin-ciple of 'heavier than air', shot across space at a speed of about four hundred miles an hour. Below him were spread out the towns with their moving pavements which carry the wayfar-ers along the streets, and the countryside, covered, as though by an immense spider's web, by the network of electric wires.

Within half an hour, Francis Bennett had reached his works at Niagara, where, after using the force of the cataracts to produce energy, he sold or hired it out to the consumers. Then, his visit over, he returned, by way of Philadelphia, Boston, and New York, to Centropolis, where his aero-car put him down about five o'clock.

The waiting room of the Earth Herald was crowded. A careful lookout was being kept for Francis Bennett to return for the daily audience he gave to his petitioners. They included the capital's acquisitive inventors, company promoters with enter-prises to suggest - all splendid, to listen to them. Among these different proposals he had to make a choice, reject the bad ones, look into the doubtful ones, give a welcome to the good ones.

He soon got rid of those who had only got useless or impracticable schemes. One of them - didn't he claim to re-vive painting, an art which had fallen into such desuetude that Millet's Angelus had just been sold for fifteen francs - thanks to the progress of colour photography invented at the end of the twentieth century by the Japanese, whose name was on every-body's lips - Aruziswa-Riochi-Nichome-Sanjukamboz-Kio-Baski-Ku? Another, hadn't he discovered the biogene bacillus which, after being introduced into the human organism, would make man immortal? This one, a chemist, hadn't he discovered

Judicious – *Sensible*
Splendid – *Grand*
Revive – *Recover*

a new substance Nihilium, of which a gram would cost only three million dollars? That one, a most daring physician, wasn't he claiming that he'd found a remedy for a cold in the head?

All these dreamers were at once shown out.

A few of the others received a better welcome, and foremost among them was a young man whose broad brow indicated a high degree of intelligence.

'Sir,' he began, 'though the number of elements used to be estimated at seventy-five, it has now been reduced to three, as no doubt you are aware?'

'Perfectly,' Francis Bennett replied.

'Well, sir, I'm on the point of reducing the three to one. If I don't run out of money I'll have succeeded in three weeks.'

'And then?'

'Then, sir, I shall really have discovered the absolute.'

'And the results of that discovery?'

'It will be to make the creation of all forms of matter easy - stone, wood, metal, fibrin . . .'

'Are you saying you're going to be able to construct a human being?'

'Completely . . . The only thing missing will be the soul!'

'Only that!' was the ironical reply of Francis Bennett, who however assigned the young fellow to the scientific editorial department of his journal.

A second inventor, using as a basis some old experiments that dated from the nineteenth century and had often been repeated since, had the idea of moving a whole city in a single block. He suggested, as a demonstration, the town of Saaf, situated fifteen miles from the sea; after conveying it on rails down to the shore, he would transform it into a seaside resort. That would add an enormous value to the ground already built on and to be built over.

Francis Bennett, attracted by this project, agreed to take a half-share in it.

'You know, sir,' said a third applicant, 'that, thanks to our solar and terrestrial accumulators and transformers, we've been able to equalise the seasons. I suggest doing even better. By converting into heat part of the energy we have at our disposal and transmitting the heat to the polar regions we can melt the ice . . .'

Demonstration –
Protest
Enormous – *Gigantic*
Disposal – *Discarding*

'Leave your plans with me,' Francis Bennett replied, 'and come back in a week.'

Finally, a fourth savant brought the news that one of the questions which had excited the whole world was about to be solved that very evening.

As is well known, a century ago a daring experiment made by Dr Nathaniel Faithburn had attracted public attention. A convinced supporter of the idea of human hibernation - the possibility of arresting the vital functions and then reawakening them after a certain time - he had decided to test the value of the method on himself. After, by a holograph will, describing the operations necessary to restore him to life a hundred years later to the day, he had exposed himself to a cold of 172 degrees centigrade (278 degrees Fahrenheit) below zero; thus reduced to a mummified state, he had been shut up in a tomb for the stated period.

Now it was exactly on that very day, 25 July 2889, that the period expired, and Francis Bennett had just received an offer to proceed in one of the rooms of the Earth Herald office with the resurrection so impatiently waited for. The public could then be kept in touch with it second by second.

The proposal was accepted, and as the operation was not to take place until ten that evening, Francis Bennett went to stretch himself out in an easy-chair in the audition room. Then, pressing a button, he was put into communication with the Central Concert.

After so busy a day, what a charm he found in the works of our greatest masters, based, as everybody knows, on a series of delicious harmonico-algebraic formulae!

The room had been darkened, and, plunged into an ecstatic half-sleep, Francis Bennett could not even see himself. But a door opened suddenly.

'Who's there?' he asked, touching a commutator placed beneath his hand.

At once, by an electric effect produced on the ether, the air became luminous.

'Oh, it's you, Doctor?' he asked.

'Myself,' replied Dr. Sam, who had come to pay his daily visit (annual subscription). 'How's it going?' 'Fine!' 'All the better . . . Let's see your tongue?'

He looked at it through a microscope.

Convince – *Persuade*
Hibernate – *Hide*
Vital – *Energetic*
Luminous – *Glowing*

'Good . . . And your pulse?'

He tested it with a pulsograph, similar to the instruments which record earthquakes.

'Splendid! . . . And your appetite?'

'Ugh!' 'Oh, your stomach! . . . It isn't going too well, your stomach! . . . It's getting old, your stomach is! . . . We'll certainly have to get you a new one!'

'We'll see!' Francis Bennett replied, 'and meantime, Doctor, you'll dine with me.'

During the meal, phonotelephotic communication had been set up with Paris. Mrs. Bennett was at her table this time, and the dinner, livened up by Dr. Sam's jokes, was delightful. Hardly was it over than:

'When do you expect to get back to Centropolis, dear Edith?' asked Francis Bennett.

'I'm going to start this moment.'

'By tube or aero-train?'

'By tube.'

'Then you'll be here?'

'At eleven fifty-nine this evening.'

'Paris time?'

'No, no! . . . Centropolis time.'

'Goodbye then, and above all don't miss the tube!'

These submarine tubes, by which one travels from Paris in two hundred and ninety-five minutes, are certainly much preferable to the aero-trains, which only manage six hundred miles an hour.

The doctor had gone, after promising to return to be present at the resurrection of his colleague Nathaniel Faithburn. Wishing to draw up his daily accounts, Francis Bennett went into his private office. An enormous operation, when it concerns an enterprise whose expenditure rises to eight hundred thousand dollars every day! Fortunately, the development of modern mechanisation has greatly facilitated this work. Helped by the piano-electric-computer, Francis Bennett soon completed his task.

It was time. Hardly had he struck the last key of the mechanical totalisator than his presence was asked for in the experimental room. He went off to it at once, and was

Delightful – *Pleasant*
Resurrection – *Revival*
Colleague – *Co-worker*

welcomed by a large cortege of scientists, who had been joined by Dr. Sam.

Nathaniel Faithburn's body is there, on the bier, placed on trestles in the centre of the room. The telephote is switched on. The whole world will be able to follow the various phases of the operation.

The coffin is opened . . . Nathaniel Faithburn's body is taken out . . . is still like a mummy, yellow hard, dry. It sounds like wood . . . It is submitted to heat . . . electricity . . . No result . . . It's hypnotised . . . It's exposed to suggestion . . . Nothing can overcome that ultracataleptic state.

'Well, Dr. Sam!' asks Francis Bennett.

The doctor leans over the body; he examines it very carefully . . . He introduces into it, by means of a hypodermic, a few drops of the famous Brown-Sequard elixir, which is once again in fashion . . . The mummy is more mummified than ever.

'Oh well,' Dr. Sam replies, 'I think the hibernation has lasted too long . . .'

'Oh!' 'And Nathaniel Faithburn is dead.'

'Dead?' 'As dead as anybody could be!'

'And how long has he been dead?'

'How long? . . .' Dr. Sam replies. But . . . a hundred years - that is to say, since he had the unhappy idea of freezing himself for pure love of science.'

'Then', Francis Bennett comments, that's a method which still needs to be perfected!'

'Perfected is the word,' replies Dr. Sam, while the scientific commission on hibernation carries away its funereal bundle.

Followed by Dr. Sam, Francis Bennett regained his room, and as he seemed very tired after so very full a day, the doctor advised him to take a bath before going to bed.

'You're quite right, Doctor . . . That will refresh me . . .'

'It will, Mr. Bennett, and if you like I'll order one on my way out . . .'

'There's no need for that, Doctor. There's always a bath all ready in the office, and I needn't even have the trouble of going out of my room to take it. Look, simply by touching this button, that bath will start moving, and you'll see it come along all by itself with the water at a temperature of sixty-five degrees!'

Cortege – *Procession*
Elixir – *Medicine*
Regain – *Recover*

Francis Bennett had just touched the button. A rumbling sound began, got louder, increased . . . Then one of the doors opened, and the bath appeared, gliding along on its rails . . . Heavens! While Dr Sam veils his face, little screams of frightened modesty arise from the bath . . .

Brought to the office by the transatlantic tube half an hour before, Mrs. Bennett was inside it.

Next day, 26 July 2889, the director of the Earth Herald recommenced his tour of twelve miles across his office. That evening, when his totalisator had been brought into action, it was at two hundred and fifty thousand dollars that it calculated the profits of that day - fifty thousand more than the day before.

A fine job, that of a journalist at the end of the twenty-ninth century!

Veils – *Coverings*

Food For Thought

Do you generally read Science Fictions? How is this one different from other fiction stories? Explain the first line of the story, 'The men of the twenty-ninth century live in a perpetual fairyland, though they do not seem to realise it' in your own words.

An Understanding

Q. 1. How do the cities in the 29th century look like?

Ans. _____

_____

_____

_____

_____

Q. 2. According to the text, how has the transportation changed by 2889?

Ans. _____

_____

_____

_____

_____

Q. 3. Why did men not make the mentioned modern inventions sooner?

Ans. _____

_____

_____

_____

_____

Q. 4. What did James Jackson discover?

Ans. _____

_____

_____

_____

_____

Saki

Born on December 18, 1870
Died on November 13, 1916
Pen name: Saki
Notable Works: *The Rise of the Russian Empire, The Woman Who Never Should, The Not So Stories, Reginald in Russia The Chronicles of Clovis* (short stories), *The Unbearable Bassington, When William Came, Beasts and Super-Beasts, The East Wing, The Interlopers, The Watched Pot, The Toys of Peace, The Storyteller, T he Open Window, Esme, The East Wing and many more.*

Early Life

Hector Hugh Munro, better known by the pen name, *Saki*, and also frequently as *H. H. Munro*, was a British writer, whose witty, mischievous and sometimes macabre stories satirised Edwardian society and culture. He is considered a master of the short story and often compared to O. Henry and Dorothy Parker. Influenced by Oscar Wilde, Lewis Carroll, and Kipling, he himself influenced A. A. Milne, Noël Coward, and P. G. Wodehouse.

Born in Akyab, Burma (now known as Myanmar), when it was still part of the British Empire, Hector Hugh Munro was the son of Charles Augustus Munro and Mary Frances Mercer.

Munro was educated at Pencarwick School in Exmouth, Devon and at Bedford School. On a few occasions, when he retired, Charles travelled with Hector and his sister to fashionable European spas and tourist resorts. In 1893, Hector followed his father into the Indian Imperial Police, where he was posted to Burma. Two years later, having contracted malaria, he resigned and returned to England.

At the start of World War I, although 43 and officially overage, Munro refused a commission and joined the British Army Royal Fusiliers as an ordinary soldier. More than once, he returned to the battlefield when officially still too sick or injured.

Literary Works and Achievements

In England, he started his career as a journalist, writing for newspapers, such as the *Westminster Gazette, Daily Express, Bystander, Morning Post* and *Outlook*. In 1900, Munro's first book appeared: *The Rise of the Russian Empire.* Shortly before the Great War, with the genre of invasion literature selling well, he also published a *What-if* novel. Beside his short stories (which were first published in newspapers, as was customary at the time, and then collected into several volumes), he wrote a full-length play, *The Watched Pot*, in collaboration with Charles Maude; two one-act plays; a historical study, *The Rise of the Russian Empire,* the only book published under his own name; a short novel, *The Unbearable Bassington; A Story of London Under the Hohenzollerns*, which was a fantasy about a future German invasion of Britain, etc. Among his popular short stories are: *The Interlopers, Gabriel-Ernest, The Toys of Peace, The Storyteller, The Open Window, The Unrest-Cure, Esme, The East Wing and many more.*

Later Years

In November 1916, Saki was sheltering in a shell crater near Beaumont-Hamel, France, when he was killed by a German sniper. After his death, his sister Ethel destroyed most of his papers and wrote her own account of their childhood.

Trivia

To recognise his contribution to English literature, **a blue plaque** has been affixed to a building in which Munro once lived on Mortimer Street in central London.

Tobermory

– Saki

IT was a chill, rain-washed afternoon of a late August day, that indefinite season when partridges are still in security or cold storage, and there is nothing to hunt - unless one is bounded on the north by the Bristol Channel, in which case one may lawfully gallop after fat red stags. Lady Blemley's house-party was not bounded on the north by the Bristol Channel, hence there was a full gathering of her guests round the tea-table on this particular afternoon. And, in spite of the blankness of the season and the triteness of the occasion, there was no trace in the company of that fatigued restlessness which means a dread of the pianola and a subdued hankering for auction bridge. The undisguised open-mouthed attention of the entire party was fixed on the homely negative personality of Mr. Cornelius Appin. Of all her guests, he was the one who had come to Lady Blemley with the vaguest reputation. Someone had said he was "clever," and he had got his invitation in the moderate expectation, on the part of his hostess, that some portion at least of his cleverness would be contributed to the general entertainment. Until tea-time that day she had been unable to discover in what direction, if any, his cleverness lay. He was neither a wit nor a croquet champion, a hypnotic force nor a begetter of amateur theatricals. Neither did his exterior suggest the sort of man in whom women are willing to pardon a generous measure of mental deficiency. He had subsided into mere Mr. Appin, and the Cornelius seemed a piece of transparent baptismal bluff. And now he was claiming to have launched on the world a discovery beside which the invention of gunpowder, of the printing press, and of steam locomotion were inconsiderable trifles. Science had made bewildering strides in many directions during recent decades, but this thing seemed to belong to the domain of miracle rather than to scientific achievement. "And do you really ask us to believe," Sir Wilfrid was saying, "that you have discovered a means for instructing animals in the art of human speech, and that dear old Tobermory has proved your first successful pupil?"

"It is a problem at which I have worked for the last seventeen years," said Mr. Appin, "but only during the last eight

Indefinite – *Unlimited*
Trite – *Commonplace*
Fatigue – *Exhaustion*
Wit – *Humour*
Amateur – *Inexpert*

or nine months have I been rewarded with glimmerings of success. Of course I have experimented with thousands of animals, but latterly only with cats, those wonderful creatures which have assimilated themselves so marvellously with our civilization while retaining all their highly developed feral instincts. Here and there among cats one comes across an outstanding superior intellect, just as one does among the ruck of human beings, and when I made the acquaintance of Tobermory a week ago I saw at once that I was in contact with a `Beyond-cat' of extraordinary intelligence. I had gone far along the road to success in recent experiments; with Tobermory, as you call him, I have reached the goal."

Mr. Appin concluded his remarkable statement in a voice which he strove to divest of a triumphant inflection. No one said "Rats," though Clovis's lips moved in a monosyllabic contortion which probably invoked those rodents of disbelief.

"And do you mean to say," asked Miss Resker, after a slight pause, "that you have taught Tobermory to say and understand easy sentences of one syllable?"

"My dear Miss Resker," said the wonder-worker patiently, "one teaches little children and savages and backward adults in that piecemeal fashion; when one has once solved the problem of making a beginning with an animal of highly developed intelligence one has no need for those halting methods. Tobermory can speak our language with perfect correctness."

This time Clovis very distinctly said, "Beyond-rats!" Sir Wilfrid was more polite, but equally sceptical. "Hadn't we better have the cat in and judge for ourselves?" suggested Lady Blemley.

Sir Wilfrid went in search of the animal, and the company settled themselves down to the languid expectation of witnessing some more or less adroit drawing room ventriloquism.

In a minute Sir Wilfrid was back in the room, his face white beneath its tan and his eyes dilated with excitement. "By Gad, it's true!"

His agitation was unmistakably genuine, and his hearers started forward in a thrill of awakened interest.

Collapsing into an armchair he continued breathlessly: "I found him dozing in the smoking room and called out to him to come for his tea. He blinked at me in his usual way, and I said, 'Come on, Toby; don't keep us waiting'; and, by

Assimilate – *Integrate*
Divest – *Strip*
Triumph – *Victory*
Halt – *Stop*
Languid – *Unhurried*

Gad! he drawled out in a most horribly natural voice that he'd come when he dashed well pleased! I nearly jumped out of my skin!"

Appin had preached to absolutely incredulous hearers; Sir Wilfred's statement carried instant conviction. A Babel-like chorus of startled exclamation arose, amid which the scientist sat mutely enjoying the first fruit of his stupendous discovery.

In the midst of the clamour Tobermory entered the room and made his way with velvet tread and studied unconcern across to the group seated round the tea-table.

A sudden hush of awkwardness and constraint fell on the company. Somehow there seemed an element of embarrassment in addressing on equal terms a domestic cat of acknowledged mental ability.

"Will you have some milk, Tobermory?" asked Lady Blemley in a rather strained voice.

"I don't mind if I do," was the response, couched in a tone of even indifference. A shiver of suppressed excitement went through the listeners, and Lady Blemley might be excused for pouring out the saucerful of milk rather unsteadily.

"I'm afraid I've spilt a good deal of it," she said apologetically.

"After all, it's not my Axminster," was Tobermory's rejoinder.

Another silence fell on the group, and then Miss Resker, in her best district-visitor manner, asked if the human language had been difficult to learn. Tobermory looked squarely at her for a moment and then fixed his gaze serenely on the middle distance. It was obvious that boring questions lay outside his scheme of life.

"What do you think of human intelligence?" asked Mavis Pellington lamely.

"Of whose intelligence in particular?" asked Tobermory coldly.

"Oh, well, mine for instance," said Mavis, with a feeble laugh.

"You put me in an embarrassing position," said Tobermory, whose tone and attitude certainly did not suggest a shred of embarrassment. "When your inclusion in this house-party was suggested, Sir Wilfrid protested that you were the most

Stupendous –
Astounding
Constraint – *Restraint*

brainless woman of his acquaintance, and that there was a wide distinction between hospitality and the care of the feeble minded. Lady Blemley replied that your lack of brain power was the precise quality which had earned you your invitation, as you were the only person she could think of who might be idiotic enough to buy their old car. You know, the one they call 'The Envy of Sisyphus,' because it goes quite nicely uphill if you push it."

Lady Blemley's protestations would have had greater effect if she had not casually suggested to Mavis only that morning that the car in question would be just the thing for her down at her Devonshire home.

Major Barfield plunged in heavily to effect a diversion.

"How about your carryings-on with the tortoise-shell puss up at the stables, eh?"

The moment he had said it everyone realized the blunder.

"One does not usually discuss these matters in public," said Tobermory frigidly. "From a slight observation of your ways since you've been in this house I should imagine you'd find it inconvenient if I were to shift the conversation on to your own little affairs."

The panic which ensued was not confined to the Major.

"Would you like to go and see if cook has got your dinner ready?" suggested Lady Blemley hurriedly, affecting to ignore the fact that it wanted at least two hours to Tobermory's dinner time. "Thanks," said Tobermory, "not quite so soon after my tea. I don't want to die of indigestion."

"Cats have nine lives, you know," said Sir Wilfrid heartily.

"Possibly", answered Tobermory; "but only one liver."

"Adelaide!" said Mrs. Cornett, "do you mean to encourage that cat to go out and gossip about us in the servants' hall?"

The panic had indeed become general. A narrow ornamental balustrade ran in front of most of the bedroom windows at the Towers, and it was recalled with dismay that this had formed a favourite promenade for Tobermory at all hours, whence he could watch the pigeons - and heaven knew what else besides. If he intended to become reminiscent in his present outspoken strain the effect would be something more than disconcerting. Mrs. Cornett, who spent much time at her toilet table, and whose complexion was reputed

Distinct – *Separate*
Precise – *Exact*
Ensue – *Follow*
Narrow – *Thin*
Dismay – *Disappointment*

to be of a nomadic though punctual disposition, looked as ill at ease as the Major. Miss Scrawen, who wrote fiercely sensuous poetry and led a blameless life, merely displayed irritation; if you are methodical and virtuous in private you don't necessarily want everyone to know it. Bertie van Tahn, who was so depraved at seventeen that he had long ago given up trying to be any worse, turned a dull shade of gardenia white, but he did not commit the error of dashing out of the room like Odo Finsberry, a young gentleman who was understood to be reading for the Church and who was possibly disturbed at the thought of scandals he might hear concerning other people. Clovis had the presence of mind to maintain a composed exterior; privately he was calculating how long it would take to procure a box of fancy mice through the agency of the Exchange and Mart as a species of hush money.

Even in a delicate situation like the present, Agnes Resker could not endure to remain too long in the background.

"Why did I ever come down here?" she asked dramatically.

Tobermory immediately accepted the opening.

"Judging by what you said to Mrs. Cornett on the croquet lawn yesterday, you were out for food. You described the Blemleys as the dullest people to stay with that you knew, but said they were clever enough to employ a first-rate cook; otherwise they'd find it difficult to get anyone to come down a second time."

"There's not a word of truth in it! I appeal to Mrs. Cornett--" exclaimed the discomfited Agnes. "Mrs. Cornett repeated your remark afterwards to Bertie van Tahn," continued Tobermory, "and said, 'That woman is a regular Hunger Marcher; she'd go anywhere for four square meals a day,' and Bertie van Tahn said--"

At this point the chronicle mercifully ceased. Tobermory had caught a glimpse of the big yellow Tom from the Rectory working his way through the shrubbery towards the stable wing. In a flash he had vanished through the open French window.

Deprave – *Corrupt*
Procure – *Acquire*
Endure – *Bear*
Cease – *Stop*
Vanish – *Disappear*

With the disappearance of his too brilliant pupil Cornelius Appin found himself beset by a hurricane of bitter upbraiding, anxious inquiry, and frightened entreaty. The responsibility for the situation lay with him, and he must prevent matters from becoming worse. Could Tobermory impart his danger-ous gift to other cats? was the first question he had to answer. It was possible, he replied, that he might have initiated his intimate friend the stable puss into his new accomplishment, but it was unlikely that his teaching could have taken a wider range as yet.

"Then," said Mrs. Cornett, "Tobermory may be a valuable cat and a great pet; but I'm sure you'll agree, Adelaide, that both he and the stable cat must be done away with without delay."

"You don't suppose I've enjoyed the last quarter of an hour, do you?" said Lady Blemley bitterly. "My husband and I are very fond of Tobermory - at least, we were before this horrible accomplishment was infused into him; but now, of course, the only thing is to have him destroyed as soon as possible."

"We can put some strychnine in the scraps he always gets at dinner time," said Sir Wilfrid, "and I will go and drown the stable cat myself. The coachman will be very sore at los-ing his pet, but I'll say a very catching form of mange has broken out in both cats and we're afraid of its spreading to the kennels."

"But my great discovery!" expostulated Mr. Appin; "after all my years of research and experiment--" "You can go and experiment on the short-horns at the farm, who are under proper control," said Mrs. Cornett, "or the elephants at the Zoological Gardens. They're said to be highly intelligent, and they have this recommendation, that they don't come creep-ing about our bedrooms and under chairs, and so forth."

An archangel ecstatically proclaiming the Millennium, and then finding that it clashed unpardonably with Henley and would have to be indefinitely postponed, could hardly have felt more crestfallen than Cornelius Appin at the recep-tion of his wonderful achievement. Public opinion, however, was against him - in fact, had the general voice been consulted on the subject it is probable that a strong minority vote would have been in favour of including him in the strychnine diet.

Defective train arrangements and a nervous desire to see matters brought to a finish prevented an immediate dispersal of the party, but dinner that evening was not a social success. Sir Wilfrid had had rather a trying time with the stable cat and

Impart – *Inform*
Infuse – *Fill*
Sore – *Painful*
Expostulate – *Remonstrate*

subsequently with the coachman. Agnes Resker ostentatiously limited her repast to a morsel of dry toast, which she bit as though it were a personal enemy; while Mavis Pellington maintained a vindictive silence throughout the meal. Lady Blemley kept up a flow of what she hoped was conversation, but her attention was fixed on the doorway. A plateful of carefully dosed fish scraps was in readiness on the sideboard, but sweets and savoury and dessert went their way, and no Tobermory appeared either in the dining room or kitchen.

The sepulchral dinner was cheerful compared with the subsequent vigil in the smoking room. Eating and drinking had at least supplied a distraction and cloak to the prevailing embarrassment. Bridge was out of the question in the general tension of nerves and tempers, and after Odo Finsberry had given a lugubrious rendering of "Melisande in the Wood" to a frigid audience, music was tacitly avoided. At eleven the servants went to bed, announcing that the small window in the pantry had been left open as usual for Tobermory's private use. The guests read steadily through the current batch of magazines, and fell back gradually on the "Badminton Library" and bound volumes of Punch. Lady Blemley made periodic visits to the pantry, returning each time with an expression of listless depression which forestalled questioning.

At two o'clock Clovis broke the dominating silence.

"He won't turn up tonight. He's probably in the local newspaper office at the present moment, dictating the first instalment of his reminiscences. Lady What's-her-name's book won't be in it. It will be the event of the day."

Having made this contribution to the general cheerfulness, Clovis went to bed. At long intervals the various members of the house party followed his example.

The servants taking round the early tea made a uniform announcement in reply to a uniform question. Tobermory had not returned.

Breakfast was, if anything, a more unpleasant function than dinner had been, but before its conclusion the situation was relieved. Tobermory's corpse was brought in from the shrubbery, where a gardener had just discovered it. From the bites on his throat and the yellow fur which coated his claws it was evident that he had fallen in unequal combat with the big Tom from the Rectory.

Vindictive – *Spiteful*
Lugubrious – *Sad*
Frigid – *Chilly*
Combat – *Battle*

By midday most of the guests had quitted the Towers, and after lunch Lady Blemley had sufficiently recovered her spirits to write an extremely nasty letter to the Rectory about the loss of her valuable pet.

Tobermory had been Appin's one successful pupil, and he was destined to have no successor. A few weeks later an elephant in the Dresden Zoological Garden, which had shown no previous signs of irritability, broke loose and killed an Englishman who had apparently been teasing it. The victim's name was variously reported in the papers as Oppin and Eppelin, but his front name was faithfully rendered Cornelius.

"If he was trying German irregular verbs on the poor beast," said Clovis, "he deserved all he got."

Food For Thought

Cornelius Appin is killed by a zoo's elephant while teaching him to speak. Do you think that he was sensible to have attempted to do so? Give reasons for your answer.

Nasty – *Horrid*
Pupil – *Apprentice*

An Understanding

Q. 1. Who was Cornelius Appin and what was the art in which he had perfected?

Ans. _____

_____

_____

_____

_____

Q. 2. Who was Tobermory? Why were the guests angry with Tobermary and what happens to Tobermory at the end?

Ans. _____

_____

_____

_____

_____

Q. 3. How did Cornelius Appin die? Why did everyone in the countryhouse party think that Appin had made a remarkable achievement?

Ans. _____

_____

_____

_____

_____

Q. 4. Do you think that Cornelius Appin had perfected in the art of teaching animals to talk? Do you really think that it was right to allow the animals to view private or confidential things?

Ans. _____

_____

_____

_____

_____

"Ministers of Grace"

~ Saki

ALthough he was scarcely yet out of his teens, the Duke of Scaw was already marked out as a personality widely differing from others of his caste and period. Not in externals; therein he conformed correctly to type. His hair was faintly reminiscent of Houbigant, and at the other end of him his shoes exhaled the right soupcon of harness room; his socks compelled one's attention without losing one's respect; and his attitude in repose had just that suggestion of Whistler's mother, so becoming in the really young. It was within that the trouble lay, if trouble it could be accounted, which marked him apart from his fellows. The Duke was religious. Not in any of the ordinary senses of the word; he took small heed of High Church or Evangelical standpoints, he stood outside of all the movements and missions and cults and crusades of the day, uncaring and uninterested. Yet in a mystical-practical way of his own, which had served him unscathed and un-shaken through the fickle years of boyhood, he was intensely and intensively religious. His family were naturally, though unobtrusively, distressed about it. "I am so afraid it may affect his bridge," said his mother.

The Duke sat in a pennyworth of chair in St. James's Park, listening to the pessimisms of Belturbet, who reviewed the existing political situation from the gloomiest of standpoints.

"Where I think you political spade-workers are so silly," said the Duke, "is in the misdirection of your efforts. You spend thousands of pounds of money, and Heaven knows how much dynamic force of brain power and personal energy, in trying to elect or displace this or that man, whereas you could gain your ends so much more simply by making use of the men as you find them. If they don't suit your purpose as they are, transform them into something more satisfactory."

"Do you refer to hypnotic suggestion?" asked Belturbet, with the air of one who is being trifled with.

"Nothing of the sort. Do you understand what I mean by the verb to koepenick? That is to say, to replace an authority by a spurious imitation that would carry just as much weight for the moment as the displaced original; the advantage, of

Scarce – *Threatened*
Repose – *Peacefulness*
Unscathed – *Unharmed*
Fickle – *Indecisive*
Spurious – *FALSE*

course, being that the koepenick replica would do what you wanted, whereas the original does what seems best in its own eyes."

"I suppose every public man has a double, if not two or three," said Belturbet; "but it would be a pretty hard task to koepenick a whole bunch of them and keep the originals out of the way."

"There have been instances in European history of highly successful koepenickery," said the Duke dreamily.

"Oh, of course, there have been False Dimitris and Perkin Warbecks, who imposed on the world for a time," assented Belturbet, "but they personated people who were dead or safely out of the way. That was a comparatively simple matter. It would be far easier to pass oneself off as dead Hannibal than as living Haldane, for instance."

"I was thinking," said the Duke, "of the most famous case of all, the angel who koepenicked King Robert of Sicily with such brilliant results. Just imagine what an advantage it would be to have angels deputizing, to use a horrible but convenient word, for Quinston and Lord Hugo Sizzle, for example. How much smoother the Parliamentary machine would work than at present!" "Now you're talking nonsense," said Belturbet; "angels don't exist nowadays, at least, not in that way, so what is the use of dragging them into a serious discussion? It's merely silly."

"If you talk to me like that I shall just do it," said the Duke.

"Do what?" asked Belturbet. There were times when his young friend's uncanny remarks rather frightened him.

"I shall summon angelic forces to take over some of the more troublesome personalities of our public life, and I shall send the ousted originals into temporary retirement in suitable animal organisms. It's not everyone who would have the knowledge or the power necessary to bring such a thing off--"

"Oh, stop that inane rubbish," said Belturbet angrily; "it's getting wearisome. Here's Quinston coming," he added, as there approached along the almost deserted path the well-known figure of a young Cabinet Minister, whose personality evoked a curious mixture of public interest and unpopularity.

"Hurry along, my dear man," said the young Duke to the Minister, who had given him a condescending nod; "your

Impose – *Levy*
Uncanny – *Weird*
Ousted – *Exiled*
Inane – *Silly*

time is running short," he continued in a provocative strain; "the whole inept crowd of you will shortly be swept away into the world's wastepaper basket."

"You poor little strawberry-leafed nonentity," said the Minister, checking himself for a moment in his stride and rolling out his words spasmodically; "who is going to sweep us away, I should like to know? The voting masses are on our side, and all the ability and administrative talent is on our side too. No power of earth or Heaven is going to move us from our place till we choose to quit it. No power of earth or--"

Belturbet saw, with bulging eyes, a sudden void where a moment earlier had been a Cabinet Minister; a void emphasized rather than relieved by the presence of a puffed-out bewildered-looking sparrow, which hopped about for a moment in a dazed fashion and then fell to a violent cheeping and scolding.

"If we could understand sparrow language," said the Duke serenely, "I fancy we should hear something infinitely worse than 'strawberry-leafed nonentity.' "

"But good Heavens, Eugene," said Belturbet hoarsely, "what has become of-- Why, there he is! How on earth did he get there?" And he pointed with a shaking finger towards a semblance of the vanished Minister, which approached once more along the unfrequented path.

The Duke laughed.

"It is Quinston to all outward appearance," he said composedly, "but I fancy you will find, on closer investigation, that it is an angel under-study of the real article."

The Angel-Quinston greeted them with a friendly smile.

"How beastly happy you two look sitting there!" he said wistfully.

"I don't suppose you'd care to change places with poor little us," replied the Duke chaffingly.

"How about poor little me?" said the Angel modestly. "I've got to run about behind the wheels of popularity, like a spotted dog behind a carriage, getting all the dust and trying to look as if I was an important part of the machine. I must seem a perfect fool to you onlookers sometimes."

"I think you are a perfect angel." said the Duke.

The Angel-that-had-been-Quinston smiled and passed on his way, pursued across the breadth of the Horse Guards

Inept – *Incompetent*
Void – *Invalid*
Pursue – *Follow*

Parade by a tiresome little sparrow that cheeped incessantly and furiously at him.

"That's only the beginning," said the Duke complacently; "I've made it operative with all of them, irrespective of parties."

Belturbet made no coherent reply; he was engaged in feeling his pulse. The Duke fixed his attention with some interest on a black swan that was swimming with haughty, stiff-necked aloofness amid the crowd of lesser water-fowl that dotted the ornamental water. For all its pride of bearing, something was evidently ruffling and enraging it; in its way it seemed as angry and amazed as the sparrow had been.

At the same moment a human figure came along the pathway. Belturbet looked up apprehensively. "Kedzon," he whispered briefly.

"An Angel-Kedzon, if I am not mistaken," said the Duke. "Look, he is talking affably to a human being. That settles it."

A shabbily dressed lounger had accosted the man who had been Viceroy in the splendid East, and who still reflected in his mien some of the cold dignity of the Himalayan snow peaks.

"Could you tell me, sir, if them white birds is storks or halbatrosses? I had an argyment--"

The cold dignity thawed at once into genial friendliness.

"Those are pelicans, my dear sir. Are you interested in birds? If you would join me in a bun and a glass of milk at the stall yonder, I could tell you some interesting things about Indian birds. Right oh! Now the hill-mynah, for instance--"

The two men disappeared in the direction of the bun stall, chatting volubly as they went, and shadowed from the other side of the railed enclosure by a black swan, whose temper seemed to have reached the limit of inarticulate rage.

Belturbet gazed in an open-mouthed wonder after the retreating couple, then transferred his attention to the infuriated swan, and finally turned with a look of scared comprehension at his young friend lolling unconcernedly in his chair. There was no longer any room to doubt what was happening. The "silly talk" had been translated into terrifying action.

"I think a prairie oyster on the top of a stiffish brandy-and-soda might save my reason," said Belturbet weakly, as he limped towards his club.

Complacent – *Content*
Coherent – *Clear*
Aloof – *Reserved*
Rage – *Anger*

It was late in the day before he could steady his nerves sufficiently to glance at the evening papers. The Parliamentary report proved significant reading, and confirmed the fears that he had been trying to shake off. Mr. Ap Dave, the Chancellor, whose lively controversial style endeared him to his supporters and embittered him, politically speaking, to his opponents, had risen in his place to make an unprovoked apology for having alluded in a recent speech to certain protesting taxpayers as "skulkers." He had realized on reflection that they were in all probability perfectly honest in their inability to understand certain legal technicalities of the new finance laws. The House had scarcely recovered from this sensation when Lord Hugo Sizzle caused a further flutter of astonishment by going out of his way to indulge in an outspoken appreciation of the fairness, loyalty, and straightforwardness not only of the Chancellor, but of all the members of the Cabinet. A wit had gravely suggested moving the adjournment of the House in view of the unexpected circumstances that had arisen.

Belturbet anxiously skimmed over a further item of news printed immediately below the Parliamentary report: "Wild cat found in an exhausted condition in Palace Yard."

"Now I wonder which of them--" he mused, and then an appalling idea came to him. "Supposing he's put them both into the same beast!" He hurriedly ordered another prairie oyster.

Belturbet was known in his club as a strictly moderate drinker; his consumption of alcoholic stimulants that day gave rise to considerable comment.

The events of the next few days were piquantly bewildering to the world at large; to Belturbet, who knew dimly what was happening, the situation was fraught with recurring alarms. The old saying that in politics it's the unexpected that always happens received a justification that it had hitherto somewhat lacked, and the epidemic of startling personal changes of front was not wholly confined to the realm of actual politics. The eminent chocolate magnate, Sadbury, whose antipathy to the Turf and everything connected with it was a matter of general knowledge, had evidently been replaced by an Angel-Sadbury, who proceeded to electrify the public by blossoming forth as an owner of race horses, giving as a reason his matured conviction

Endear – *Commend*
Embitter – *Disillusion*
Allude – *Mention*
Fraught – *Tense*

that the sport was, after all, one which gave healthy open-air recreation to large numbers of people drawn from all classes of the community, and incidentally stimulated the important industry of horse breeding. His colours, chocolate and cream hoops spangled with pink stars, promised to become as popular as any on the Turf. At the same time, in order to give effect to his condemnation of the evils resulting from the spread of the gambling habit among wage-earning classes, who lived for the most part from hand to mouth, he suppressed all betting news and tipsters' forecasts in the popular evening paper that was under his control. His action received instant recognition and support from the Angel-proprietor of the *Evening Views*, the principal rival evening half penny paper, who forthwith issued an ukase decreeing a similar ban on betting news, and in a short while the regular evening Press was purged of all mention of starting prices and probable winners. A considerable drop in the circulation of all these papers was the immediate result, accompanied, of course, by a falling off in advertisement value, while a crop of special betting broadsheets sprang up to supply the newly created want. Under their influence the betting habit became if anything rather more widely diffused than before. The Duke had possibly overlooked the futility of koepenicking the leaders of the nation with excellently intentioned angel under-studies, while leaving the mass of the people in its original condition.

Further sensation and dislocation was caused in the Press world by the sudden and dramatic repprochement which took place between the Angel-Editor of the *Scrutator* and the Angel-Editor of the *Anglian Review*, who not only ceased to criticize and disparage the tone and tendencies of each other's publication, but agreed to exchange editorships for alternating periods. Here again public support was not on the side of the angels; constant readers of the *Scrutator* complained bitterly of the strong meat which was thrust upon them at fitful intervals in place of the almost vegetarian diet to which they had become confidently accustomed; even those who were not mentally averse to strong meat as a separate course were pardonably annoyed at being supplied with it in the pages of the *Scrutator*. To be suddenly confronted with a pungent herring salad when one had attuned oneself to tea and toast, or to discover a richly truffled segment of *pate de foie* dissembled in a bowl of bread and milk, would be

Rival – *Competing*
Cease – *Stop*
Averse – *Opposed*
Pungent – *Strong*

an experience that might upset the equanimity of the most placidly disposed mortal. An equally vehement outcry arose from the regular subscribers of the *Anglian Review*, who protested against being served from time to time with literary fare which no young person of sixteen could possibly want to devour in secret. To take infinite precautions, they complained, against the juvenile perusal of such eminently innocuous literature was like reading the Riot Act on an uninhabited island. Both reviews suffered a serious falling-off in circulation and influence. Peace hath its devastations as well as war.

The wives of noted public men formed another element of discomfiture which the young Duke had almost entirely left out of his calculations. It is sufficiently embarrassing to keep abreast of the possible wobblings and veerings-round of a human husband, who, from the strength or weakness of his personal character, may leap over or slip through the barriers which divide the parties; for this reason a merciful politician usually marries late in life, when he has definitely made up his mind on which side he wishes his wife to be socially valuable. But these trials were nothing compared to the bewilderment caused by the Angel-husbands who seemed in some cases to have revolutionized their outlook on life in the interval between breakfast and dinner, without premonition or preparation of any kind, and apparently without realizing the least need for subsequent explanation. The temporary peace which brooded over the Parliamentary situation was by no means reproduced in the home circles of the leading statesmen and politicians. It had been frequently and extensively remarked of Mrs. Exe that she would try the patience of an angel; now the tables were reversed, and she unwittingly had an opportunity for discovering that the capacity for exasperating behaviour was not all on one side.

And then, with the introduction of the Navy Estimates, Parliamentary peace suddenly dissolved. It was the old quarrel between Ministers and the Opposition as to the adequacy or the reverse of the Government's naval programme. The Angel-Quinston and the Angel-Hugo-Sizzle contrived to keep the debates free from personalities and pinpricks, but an enormous sensation was created when the elegant lackadaisical Halfan Halfour threatened to bring up fifty thousand

Vehement –
Passionate
Devour – *Consume*
Discomfiture –
Embarrassment
Contrive – *Plan*

stalwarts to wreck the House if the Estimates were not forthwith revised on a Two-Power basis. It was a memorable scene when he rose in his place, in response to the scandalized shouts of his opponents, and thundered forth, "Gentlemen, I glory in the name of Apache."

Belturbet, who had made several fruitless attempts to ring up his young friend since the fateful morning in St. James's Park, ran him to earth one afternoon at his club, smooth and spruce and unruffled as ever.

"Tell me, what on earth have you turned Cocksley Coxon into?" Belturbet asked anxiously, mentioning the name of one of the pillars of unorthodoxy in the Anglican Church. "I don't fancy he believes in angels, and if he finds an angel preaching orthodox sermons from his pulpit while he's been turned into a fox-terrier, he'll develop rabies in less than no time."

"I rather think it was a fox-terrier," said the Duke lazily.

Belturbet groaned heavily, and sank into a chair.

"Look here, Eugene," he whispered hoarsely, having first looked well round to see that no one was within hearing range, "you've got to stop it. Consols are jumping up and down like bronchos, and that speech of Halfour's in the House last night has simply startled everybody out of their wits. And then on the top if it, Thistlebery--"

"What has he been saying?" asked the Duke quickly.

"Nothing. That's just what's so disturbing. Everyone thought it was simply inevitable that he should come out with a great epoch-making speech at this juncture, and I've just seen on the tape that he has refused to address any meetings at present, giving as a reason his opinion that something more than mere speech-making was wanted."

The young Duke said nothing, but his eyes shone with quiet exultation.

"It's so unlike Thistlebery," continued Belturbet; "at least," he said suspiciously, "it's unlike the real Thistlebery--"

"The real Thistlebery is flying about somewhere as a vocally industrious lapwing," said the Duke calmly; "I expect great things of the Angel-Thistlebery," he added.

At this moment there was a magnetic stampede of members towards the lobby, where the tape machines were ticking out some news of more than ordinary import.

Spruce – *Neat*
Inevitable – *Unavoidable*
Exultation – *Happiness*
Stampede – *Rush*

"*Coup d'etat* in the North. Thistlebery seizes Edinburgh Castle. Threatens civil war unless Government expands naval programme."

In the babel which ensued Belturbet lost sight of his young friend. For the best part of the afternoon he searched one likely haunt after another, spurred on by the sensational posters which the evening papers were displaying broadcast over the West End. General Baden-Baden mobilizes Boy-Scouts. Another *coup d'etat* feared. Is Windsor Castle safe?" This was one of the earlier posters, and was followed by one of even more sinister purport: "Will the Test match have to be postponed?" It was this disquietening question which brought home the real seriousness of the situation to the London public, and made people wonder whether one might not pay too high a price for the advantages of party government. Belturbet, questing round in the hope of finding the originator of the trouble, with a vague idea of being able to induce him to restore matters to their normal human footing, came across an elderly club acquaintance who dabbled extensively in some of the more sensitive market securities. He was pale with indignation, and his pallor deepened as a breathless newsboy dashed past with a poster inscribed: "Premier's constituency harried by moss-troopers. Halfour sends encouraging telegram to rioters. Letchworth Garden City threatens reprisals. Foreigners taking refuge in Embassies and National Liberal Club."

"This is devils' work!" he said angrily.

Belturbet knew otherwise.

At the bottom of St. James's Street a newspaper motor-cart, which had just come rapidly along Pall Mall, was surrounded by a knot of eagerly talking people, and for the first time that afternoon Belturbet heard expressions of relief and congratulation.

It displayed a placard with the welcome announcement: "Crisis ended. Government gives way. Important expansion of naval programme."

There seemed to be no immediate necessity for pursuing the quest of the errant Duke, and Belturbet turned to make his way homeward through St. James's Park. His mind, attuned to the alarms and excursions of the afternoon, became dimly aware that some excitement of a detached nature was going on around him. In spite of the political ferment which reigned in the streets, quite a large crowd had gathered to watch the unfolding of a tragedy that had taken place on the shore of

Ensue – *Follow*
Sinister – *Ominous*
Pallor – *Paleness*
Quest – *Mission*
Attune – *Adjust*

the ornamental water. A large black swan, which had recently shown signs of a savage and dangerous disposition, had suddenly attacked a young gentleman who was walking by the water's edge, dragged him down under the surface, and drowned him before anyone could come to his assistance. At the moment when Belturbet arrived on the spot several park-keepers were engaged in lifting the corpse into a punt. Belturbet stooped to pick up a hat that lay near the scene of the struggle. It was a smart soft felt hat, faintly reminiscent of Houbigant.

More than a month elapsed before Belturbet had sufficiently recovered from his attack of nervous prostration to take an interest once more in what was going on in the world of politics. The Parliamentary Session was still in full swing, and a General Election was looming in the near future. He called for a batch of morning papers and skimmed rapidly through the speeches of the Chancellor, Quinston, and other Ministerial leaders, as well as those of the principal Opposition champions, and then sank back in his chair with a sigh of relief. Evidently the spell had ceased to act after the tragedy which had overtaken its invoker. There was no trace of angel anywhere.

Tragedy – *Disaster*
Punt – *Boot*
Reminiscent – *Significant*

Food For Thought

Why did Belturbet heave a sigh of relief after going through the speeches of the Chancellor, Quinston and other ministerial leaders? 'Evidenty, the spell had ceased to act after the tragedy which had overtaken its invoker.' What was the spell and what was the tragedy all about? Explain.

An Understanding

Q. 1. How was the Duke of Scaw different from others of his caste and period?

Ans. _____

_____

_____

_____

_____

Q. 2. What is the plot of the story in brief? Why has the author, chosen this title, 'Ministers of Grace' for the story?

Ans. _____

_____

_____

_____

_____

Q. 3. What did Belturbet come across while he was on his way home-ward through St. James's Park? What had happened there?

Ans. _____

_____

_____

_____

_____

Q. 4. What had happened to Belturbet? Why did he suffer from an attack of nervous prostration?

Ans. _____

_____

_____

_____

_____

John William Polidori

Born on September 7, 1795
Died on August 24, 1821 (aged 25)
Genres: Vampire, Horror
Notable Works: *The Vampyre, On the Punishment of Death (1816), An Essay Upon the Source of Positive Pleasure, The Stress of Her Regard, The Fall of the Angels* and many more.

Early Life

John William Polidori was an English writer and physician of Italian descent. He is known for his associations with the Romantic movement and credited by some as the creator of the vampire genre of fantasy fiction. His most successful work was the 1819 short story, *The Vampyre*, one of the first vampire stories in English.

John William Polidori was born in 1795 in London, England, the eldest son of Gaetano Polidori, an Italian political émigré scholar, and Anna Maria Pierce, an English governess. He had three brothers and four sisters.

Polidori was one of the earliest pupils at recently established Ampleforth College from 1804, and in 1810 went up to the University of Edinburgh, where he wrote a thesis on sleepwalking and received his degree as a doctor of medicine on August 1, 1815 at the age of 19.

Literary Works and Achievements

In 1816, Dr. Polidori entered Lord Byron's service as his personal physician, and accompanied Byron on a trip through Europe. Publisher John Murray offered Polidori 500 English pounds to keep a diary of their travels, which Polidori's nephew, William Michael Rossetti later edited. *The Vampyre* is the first vampire story published in English.

Polidori's long, Byron-influenced theological poem, *The Fall of the Angels*, was published anonymously in 1821.

The Stress of Her Regard (1989), in which Polidori does not write about vampires, but becomes directly involved with them. In Powers' sequel (of sorts), *Hide Me Among the Graves* (2012), Polidori is a vampire and a central villain menacing the novel's protagonists, his nieces and nephews in the Rossetti family. Paul West's novel, *Lord Byron's Doctor* (1989) is a recreation, and ribald fictionalisation, of Polidori's diaries. West depicts him as a literary groupie whose attempts to emulate Byron eventually unhinge and destroy him.Polidori is a central character in the novel, *The Merciful Women* (or Las Piadosas in the original Argentine edition) by Federico Andahazi.

Polidori is also a central character in the novel, *Gothic Romance* and *The Casebook of Victor Frankenstein.*

Some of the well-known works of John William Polidori are: *A Medical Inaugural Dissertation which deals with the disease called Oneirodynia, for the degree of Medical Doctor, Edinburgh (1815), On the Punishment of Death (1816), An Essay Upon the Source of Positive Pleasure (1818), The Vampyre: A Tale (1819), Ernestus Berchtold; or, The Modern Oedipus: A Tale (1819), The Diary of Dr. John William Polidori (1816), Ximenes, The Wreath and Other Poems (1819), The Fall of the Angels: A Sacred Poem (1821), etc.*

Later Years

Polidori died in London on August 24, 1821, weighed down by depression and gambling debts. Despite strong evidence that he committed suicide by means of prussic acid (cyanide), the coroner gave a verdict of death by natural causes.

The Vampyre

~ John Polidori

IT happened that in the midst of the dissipations attendant upon London winter, there appeared at the various parties of the leaders of the ton a nobleman more remarkable for his singularities, than his rank. He gazed upon the mirth around him, as if he could not participate therein. Apparently, the light laughter of the fair only attracted his attention, that he might by a look quell it and throw fear into those breasts where thoughtlessness reigned. Those who felt this sensation of awe, could not explain whence it arose: some attributed it to the dead grey eye, which, fixing upon the object's face, did not seem to penetrate, and at one glance to pierce through to the inward workings of the heart; but fell upon the cheek with a leaden ray that weighed upon the skin it could not pass. His peculiarities caused him to be invited to every house; all wished to see him, and those who had been accustomed to violent excitement, and now felt the weight of ennui, were pleased at having something in their presence capable of engaging their attention. Inspite of the deadly hue of his face, which never gained a wanner tint, either from the blush of modesty, or from the strong emotion of passion, though its form and outline were beautiful, many of the female hunters after notoriety attempted to win his attentions, and gain, at least, some marks of what they might term affection: Lady Mercer, who had been the mockery of every monster shewn in drawing rooms since her marriage, threw herself in his way, and did all but put on the dress of a mountebank, to attract his notice - though in vain; - when she stood before him, though his eyes were apparently fixed upon hers, still it seemed as if they were unperceived; even her unappalled impudence was baffled, and she left the field. But though the common adultress could not influence even the guidance of his eyes, it was not that the female sex was indifferent to him, yet such was the apparent caution with which he spoke to the virtuous wife and innocent daughter, that few knew he ever addressed himself to females. He had, however, the reputation of a winning tongue; and whether it was that it even overcame the dread of his singular character, or that they were moved by his apparent hatred of vice, he was as often among those females

Mirth – *Amusement*
Quell – *Crush*
Peculiar – *Weird*
Dread – *Terror*

who form the boast of their sex from their domestic virtues, as among those who sully it by their vices.

About the same time, there came to London a young gentleman of the name of Aubrey. He was an orphan left with an only sister in the possession of great wealth, by parents who died while he was yet in childhood. Left also to himself by guardians, who thought it their duty merely to take care of his fortune, while they relinquished the more important charge of his mind to the care of mercenary subalterns, he cultivated more his imagination than his judgment. He had, hence, that high romantic feeling of honour and candour, which daily ruins so many milliners' apprentices. He believed all to sympathize with virtue, and thought that vice was thrown in by Providence merely for the picturesque effect of the scene, as we see in romances. He thought that the misery of a cottage merely consisted in the vesting of clothes, which were as warm, but which were better adapted to the painter's eye by their irregular folds and various coloured patches. He thought, in fine, that the dreams of poets were the realities of life. He was handsome, frank, and rich: for these reasons, upon his entering into the gay circles, many mothers surrounded him, striving which should describe with least truth their languishing or romping favourites; the daughters at the same time, by their brightening countenances when he approached, and by their sparkling eyes, when he opened his lips, soon led him into false notions of his talents and his merit. Attached as he was to the romance of his solitary hours, he was startled at finding, that, except in the tallow and wax candles that flickered, not from the presence of a ghost, but from want of snuffing, there was no foundation in real life for any of that congeries of pleasing pictures and descriptions contained in those volumes, from which he had formed his study. Finding, however, some compensation in his gratified vanity, he was about to relinquish his dreams, when the extraordinary being we have above described, crossed him in his career.

He watched him; and the very impossibility of forming an idea of the character of a man entirely absorbed in himself, who gave few other signs of his observation of external objects, than the tacit assent to their existence, implied by the avoidance of their contact: allowing his imagination to picture every thing that flattered its propensity to extravagant ideas, he soon formed this object into the hero of a romance,

Sully – *Smear*
Relinquish – *Give up*
Countenance – *Tolerate*
Notion – *Idea*

and determined to observe the offspring of his fancy, rather than the person before him. He became acquainted with him, paid him attention, and so far advanced upon his notice, that his presence was always recognised. He gradually learnt that Lord Ruthven's affairs were embarrassed, and soon found, from the notes of preparation in Street, that he was about to travel. Desirous of gaining some information respecting this singular character, who, till now, had only whetted his curiosity, he hinted to his guardians, that it was time for him to perform the tour, which for many generations has been thought necessary to enable the young to take some rapid steps in the career of vice towards putting themselves upon an equality with the aged, and not allowing them to appear as if fallen from the skies, whenever scandalous intrigues are mentioned as the subjects of pleasantry or of praise, according to the degree of skill shewn in carrying them on. They consented and Aubrey immediately mentioning his intentions to Lord Ruthven, was surprised to receive from him a proposal to join him. Flattered such a mark of esteem from him, who, apparently, had nothing in common with other men, he gladly accepted it, and in a few days they had passed the circling waters.

Hitherto, Aubrey had had no opportunity of studying Lord Ruthven's character, and now he found, that, though many more of his actions were exposed to his view, the results offered different conclusions from the apparent motives to his conduct. His companion was profuse in his liberality - the idle, the vagabond, and the beggar, received from his hand more than enough to relieve their immediate wants. But Aubrey could not avoid remarking that it was not upon the virtuous, reduced to indigence by the misfortunes attendant even upon virtue, that he bestowed his alms - these were sent from the door with hardly suppressed sneers; but when the profligate came to ask something, not to relieve his wants, but to allow him to wallow in his lust, to sink him still deeper in his iniquity, he was sent away with rich charity. This was, however, attributed by him to the greater importunity of the vicious, which generally prevails over the retiring bashfulness of the virtuous indigent. There was one circumstance about the charity of his Lordship, which was still more impressed upon his mind. All those upon whom it was bestowed, inevitably found that there was a curse upon it, for they were all either led to the scaffold, or sunk to the lowest and the most

Whet – *Sharpen*
Intrigue
– *Conspiracy*
Profuse – *Plentiful*
Sneer – *Scorn*

abject misery. At Brussels and other towns through which they passed, Aubrey was surprised at the apparent eagerness with which his companion sought for the centres of all fashionable vice; there he entered into all the spirit of the faro table. He betted and always gambled with success, except where the known sharper was his antagonist, and then he lost even more than he gained; but it was always with the same unchanging face, with which he generally watched the society around. It was not, however, so when he encountered the rash youthful novice, or the luckless father of a numerous family; then his very wish seemed fortune's law - this apparent abstractedness of mind was laid aside, and his eyes sparkled with more fire than that of the cat whilst dallying with the half-dead mouse. In every town, he left the formerly affluent youth, torn from the circle he adorned, cursing, in the solitude of a dungeon, the fate that had drawn him within the reach of this fiend; whilst many a father sat frantic, amidst the speaking looks of mute hungry children, without a single farthing of his late immense wealth, wherewith to buy even sufficient to satisfy their present craving. Yet he took no money from the gambling table; but immediately lost, to the ruiner of many, the last gilder he had just snatched from the convulsive grasp of the innocent. This might but be the result of a certain degree of knowledge, which was not, however, capable of combating the cunning of the more experienced. Aubrey often wished to represent this to his friend, and beg him to resign that charity and pleasure which proved the ruin of all, and did not tend to his own profit; but he delayed it - for each day he hoped his friend would give him some opportunity of speaking frankly and openly to him; however, this never occurred. Lord Ruthven in his carriage, and amidst the various wild and rich scenes of nature, was always the same. His eye spoke less than his lip; and though Aubrey was near the object of his curiosity, he obtained no greater gratification from it than the constant excitement of vainly wishing to break that mystery, which to his exalted imagination began to assume the appearance of something supernatural.

They soon arrived at Rome, and Aubrey for a time lost sight of his companion; he left him in daily attendance upon the morning circle of an Italian countess, whilst he went in search of the memorials of another almost deserted city. Whilst he was thus engaged, letters arrived from England,

Abject – *Hopeless*
Novice – *Beginner*
Affluent – *Wealthy*
Gratification – *Satisfaction*

which he opened with eager impatience; the first was from his sister, breathing nothing but affection; the others were from his guardians, the latter astonished him; if it had before entered into his imagination that there was an evil power resident in his companion these seemed to give him almost sufficient reason for the belief. His guardians insisted upon his immediately leaving his friend, and urged that his character was dreadfully vicious, for that the possession of irresistible powers of seduction, rendered his licentious habits more dangerous to society. It had been discovered, that his contempt for the adultress had not originated in hatred of her character; but that he had required, to enhance his gratification, that his victim, the partner of his guilt, should be hurled from the pinnacle of unsullied virtue, down to the lowest abyss of infamy and degradation. In fine, that all those females whom he had sought, apparently on account of their virtue, had, since his departure, thrown even the mask aside, and had not scrupled to expose the whole deformity of their vices to the public gaze.

Aubrey determined upon leaving one, whose character had not shown a single bright point on which to rest the eye. He resolved to invent some plausible pretext for abandoning him altogether, purposing, in the meanwhile, to watch him more closely, and to let no slight circumstances pass by unnoticed. He entered into the same circle, and soon perceived, that his Lordship was endeavouring to work upon the inexperience of the daughter of the lady whose house he chiefly frequented. In Italy, it is seldom that an unmarried female is met with in society; he was therefore obliged to carry on his plans in secret; but Aubrey's eye followed him in all his windings, and soon discovered that an assignation had been appointed, which would most likely end in the ruin of an innocent, though thoughtless girl. Losing no time, he entered the apartment of Lord Ruthven, and abruptly asked him his intentions with respect to the lady, informing him at the same time that he was aware of his being about to meet her that very night. Lord Ruthven answered that his intentions were such as he supposed all would have upon such an occasion; and upon being pressed whether he intended to marry her, merely laughed. Aubrey retired; and, immediately writing a note, to say, that from that moment he must decline accompanying his Lordship in the remainder of their proposed tour, he ordered

Latter – *Concluding*
Vicious – *Brutal*
Licentious –
Abandoned
Decline – *Reject*

his servant to seek other apartments, and calling upon the mother of the lady informed her of all he knew, not only with regard to her daughter, but also concerning the character of his Lordship. The assignation was prevented. Lord Ruthven next day merely sent his servant to notify his complete assent to a separation; but did not hint any suspicion of his plans having been foiled by Aubrey's interposition.

Having left Rome, Aubrey directed his steps towards Greece, and crossing the Peninsula, soon found himself at Athens. He then fixed residence in the house of a Greek; and soon occupied himself in tracing the faded records of ancient glory upon monuments that apparently, ashamed of chronicling the deeds of freemen only before slaves, had hidden themselves beneath the sheltering soil or many coloured lichen. Under the same roof as himself, existed a being, so beautiful and delicate, that she might have formed the model for a painter, wishing to portray on canvas the promised hope of the faithful in Mahomet's paradise, save that her eyes spoke too much mind for anyone to think she could belong to those who had no souls. As she danced upon the plain, or tripped along the mountain's side, one would have thought the gazelle a poor type of her beauties; for who would have exchanged her eye, apparently the eye of animated nature, for that sleepy luxurious look of the animal suited but to the taste of an epicure. The light step of Ianthe often accompanied Aubrey in his search after antiquities, and often would the unconscious girl, engaged in the pursuit of a Kashmere butterfly, show the whole beauty of her form, boating as it were upon the wind, to the eager gaze of him, who forgot the letters he had just decyphered upon an almost effaced tablet, in the contemplation of her sylph-like figure. Often would her tresses falling, as she flitted around, exhibit in the sun's ray such delicately brilliant and swiftly fading hues, as might well excuse the forgetfulness of the antiquary, who let escape from his mind the very object he had before thought of vital importance to the proper interpretation of a passage in Pausanias. But why attempt to describe charms which all feel, but none can appreciate? It was innocence, youth, and beauty, unaffected by crowded drawing rooms and stifling balls. Whilst he drew those remains of which he wished to preserve a memorial for his future hours, she would stand by, and watch the magic effects of his pencil, in tracing the scenes

Assent – *Agree*
Deed – *Action*
Epicure – *Gourmet*

of her native place; she would then describe to him the circling dance upon the open plain, would paint to him in all the glowing colours of youthful memory, the marriage pomp she remembered viewing in her infancy; and then, turning to subjects that had evidently made a greater impression upon her mind, would tell him all the supernatural tales of her nurse. Her earnestness and apparent belief of what she narrated, excited the interest even of Aubrey; and often as she told him the tale of the living vampyre, who had passed years amidst his friends, and dearest ties, forced every year, by feeding upon the life of a lovely female to prolong his existence for the ensuing months, his blood would run cold, whilst he attempted to laugh her out of such idle and horrible fantasies; but Ianthe cited to him the names of old men, who had at last detected one living among themselves, after several of their near relatives and children had been found marked with the stamp of the fiend's appetite; and when she found him so incredulous, she begged of him to believe her, for it had been remarked, that those who had dared to question their existence, always had some proof given, which obliged them, with grief and heartbreaking, to confess it was true. She detailed to him the traditional appearance of these monsters, and his horror was increased by hearing a pretty accurate description of Lord Ruthven; he, however, still persisted in persuading her, that there could be no truth in her fears, though at the same time he wondered at the many coincidences which had all tended to excite a belief in the supernatural power of Lord Ruthven.

Aubrey began to attach himself more and more to Ianthe; her innocence, so contrasted with all the affected virtues of the women among whom he had sought for his vision of romance, won his heart and while he ridiculed the idea of a young man of English habits, marrying an uneducated Greek girl, still he found himself more and more attached to the almost fairy form before him. He would tear himself at times from her, and, forming a plan for some antiquarian research, would depart, determined not to return until his object was attained; but he always found it impossible to fix his attention upon the ruins around him, whilst in his mind he retained an image that seemed alone the rightful possessor of his thoughts. Ianthe was unconscious of his love, and was ever the same frank infantile being he had first known. She always seemed to part from him with reluctance; but it was because

Pomp – *Spectacle*
Prolong – *Extend*
Virtue – *Asset*
Reluctance –
Unwillingness

she had no longer anyone with whom she could visit her favourite haunts, whilst her guardian was occupied in sketching or uncovering some fragment which had yet escaped the destructive hand of time. She had appealed to her parents on the subject of Vampyres, and they both, with several present, affirmed their existence, pale with horror at the very name. Soon after, Aubrey determined to proceed upon one of his excursions, which was to detain him for a few hours; when they heard the name of the place, they all at once begged of him not to return at night, as he must necessarily pass through a wood, where no Greek would ever remain, after the day had closed, upon any consideration. They described it as the resort of the vampyres in their nocturnal orgies and denounced the most heavy evils as impending upon him who dared to cross their path. Aubrey made light of their representations, and tried to laugh them out of the idea; but when he saw them shudder at his daring thus to mock a superior, infernal power, the very name of which apparently made their blood freeze, he was silent.

Next morning Aubrey set off upon his excursion unattended; he was surprised to observe the melancholy face of his host, and was concerned to find that his words, mocking the belief of those horrible fiends, had inspired them with such terror. When he was about to depart, Ianthe came to the side of his horse, and earnestly begged of him to return, night allowed the power of these beings to be put in action; he promised. He was, however, so occupied in his research, that he did not perceive that daylight would soon end, and that in the horizon there was one of those specks which, in the warmer climates, so rapidly gather into a tremendous mass, and pour all their rage upon the devoted country. He at last, however, mounted his horse, determined to make up by speed for his delay, but it was too late. Twilight, in these southern climates, is almost unknown; immediately the sun sets, night begins, and ere he had advanced far, the power of the storm was above - its echoing thunders had scarcely an interval of rest - its thick heavy rain forced its way through the canopying foliage, whilst the blue forked lightning seemed to fall and radiate at his very feet. Suddenly his horse took fright, and he was carried with dreadful rapidity through the entangled forest. The animal at last, through fatigue, stopped, and he found, by the glare of lightning, that he was in the

Haunt – *Trouble*
Excursion – *Tour*
Twilight – *Dusk*
Fatigue – *Exhaustion*

neighbourhood of a hovel that hardly lifted itself up from the masses of dead leaves and brushwood which surrounded it. Dismounting, he approached, hoping to find someone to guide him to the town, or at least trusting to obtain shelter from the pelting of the storm. As he approached, the thunders, for a moment silent, allowed him to hear the dreadful shrieks of a woman mingling with the stifled, exultant mockery of a laugh, continued in one almost unbroken sound. He was startled; but roused by the thunder which again rolled over his head, he, with a sudden effort, forced open the door of the hut. He found himself in utter darkness. The sound, however, guided him. He was apparently unperceived; for though he called, still the sounds continued, and no notice was taken of him. He found himself in contact with someone, whom he immediately seized; when a voice cried, "Again baffled!" to which a loud laugh succeeded; and he felt himself grappled by one whose strength seemed superhuman; determined to sell his life as dearly as he could, he struggled; but it was in vain. He was lifted from his feet and hurled with enormous force against the ground. His enemy threw himself upon him, and kneeling upon his breast, had placed his hands upon his throat when the glare of many torches penetrating through the hole that gave light in the day, disturbed him. He instantly rose, and leaving his prey, rushed through the door, and in a moment the crashing of branches, as he broke through the wood, was no longer heard. The storm was now still; and Aubrey, incapable of moving, was soon heard by those without. They entered; the light of their torches fell upon mud walls, and the thatch loaded on every individual straw with heavy flakes of soot. At the desire of Aubrey they searched for her who had attracted him by her cries; he was again left in darkness; but what was his horror, when the light of the torches once more burst upon him, to perceive the airy form of his fair conductress brought in a lifeless corpse. He shut his eyes, hoping that it was but a vision arising from his disturbed imagination; but he again saw the same form, when he unclosed them, stretched by his side. There was no colour upon her cheek, not even upon her lip; yet there was a stillness about her face that seemed almost as attaching as the life that once dwelt there. Upon her neck and breast was blood, and upon her throat were the marks of teeth having opened the vein. To this the men pointed, crying, simultaneously struck

Dreadful – *Terrible*
Seize – *Grab*
Prey – *Victim*
Corpse – *Dead body*

with horror, "A Vampyre! a Vampyre!" A litter was quickly formed, and Aubrey was laid by the side of her who had lately been to him the object of so many bright and fairy visions, now fallen; with the flower of life that had died within her. He knew not what his thoughts were - his mind was benumbed and seemed to shun reflection and take refuge in vacancy; he held almost unconsciously in his hand a naked dagger of a particular construction, which had been found in the hut. They were soon met by different parties who had been engaged in the search of her whom a mother had missed. Their lamentable cries as they approached the city, forewarned the parents of some dreadful catastrophe. To describe their grief would be impossible; but when they ascertained the cause of their child's death, they looked at Aubrey and pointed to the corpse. They were inconsolable; both died brokenhearted.

Aubrey being put to bed was seized with a most violent fever, and was often delirious; in these intervals he would call upon Lord Ruthven and upon Ianthe - by some unaccountable combination he seemed to beg of his former companion to spare the being he loved. At other times he would imprecate maledictions upon his head, and curse him as her destroyer. Lord Ruthven chanced at this time to arrive at Athens, and from whatever motive, upon hearing of the state of Aubrey, immediately placed himself in the same house, and became his constant attendant. When the latter recovered from his delirium, he was horrified and startled at the sight of him whose image he had now combined with that of a Vampyre; but Lord Ruthven, by his kind words, implying almost repentance for the fault that had caused their separation, and still more by the attention, anxiety, and care which he showed, soon reconciled him to his presence. His lordship seemed quite changed; he no longer appeared that apathetic being who had so astonished Aubrey; but as soon as his convalescence began to be rapid, he again gradually retired into the same state of mind, and Aubrey perceived no difference from the former man, except that at times he was surprised to meet his gaze fixed intently upon him, with a smile of malicious exultation playing upon his lips. He knew not why, but this smile haunted him. During the last stage of the invalid's recovery, Lord Ruthven was apparently engaged in watching the tideless waves raised by the cooling breeze, or in marking the progress of those

Shun – *Avoid*
Lament – *Mourn*
Apathetic – *Uninterested*
Malicious – *Hateful*

orbs, circling, like our world, the moveless sun; indeed he appeared to wish to avoid the eyes of all.

Aubrey's mind, by this shock, was much weakened, and that elasticity of spirit which had once so distinguished him now seemed to have fled forever. He was now as much a lover of solitude and silence as Lord Ruthven; but much as he wished for solitude, his mind could not find it in the neighbourhood of Athens; if he sought it amidst the ruins he had formerly frequented, Ianthe's form stood by his side; if he sought it in the woods, her light step would appear wandering amidst the underwood, in quest of the modest violet; then suddenly turning round, would show, to his wild imagination, her pale face and wounded throat, with a meek smile upon her lips. He determined to fly scenes, every feature of which created such bitter associations in his mind. He proposed to Lord Ruthven, to whom he held himself bound by the tender care he had taken of him during his illness, that they should visit those parts of Greece neither had yet seen. They travelled in every direction, and sought every spot to which a recollection could be attached, but though they thus hastened from place to place, yet they seemed not to heed what they gazed upon. They heard much of robbers, but they gradually began to slight these reports, which they imagined were only the invention of individuals, whose interest it was to excite the generosity of those whom they defended from pretended dangers. In consequence of thus neglecting the advice of the inhabitants, on one occasion they travelled with only a few guards, more to serve as guides than as a defence. Upon entering, however, a narrow defile, at the bottom of which was the bed of a torrent, with large masses of rock brought down from the neighbouring precipices, they had reason to repent their negligence; for scarcely were the whole of the party engaged in the narrow pass, when they were startled by the whistling of bullets close to their heads, and by the echoed report of several guns. In an instant their guards had left them, and, placing themselves behind rocks, had begun to fire in the direction whence the report came. Lord Ruthven and Aubrey, imitating their example, retired for a moment behind the sheltering turn of the defile, but ashamed of being thus detained by a foe, who with insulting shouts bade them advance, and being exposed to unresisting slaughter, if any of the robbers should climb above and take them in the rear, they determined

Meek – *Humble*
Heed – *Observe*
Defile – *Taint*
Repent – *Regret*
Foe – *Enemy*

at once to rush forward in search of the enemy. Hardly had they lost the shelter of rock, when Lord Ruthven received a shot in the shoulder, which brought him to the ground. Aubrey hastened to his assistance; and, no longer heeding the contest or his own peril, was soon surprised by seeing the robbers' faces around him - his guards having, upon Lord Ruthven's being wounded, immediately thrown up their arms and surrendered.

By promises of great reward, Aubrey soon induced them to convey his wounded friend to a neighbouring cabin; and having agreed upon a ransom, he was no more disturbed by their presence - they being content merely to guard the entrance till their comrade should return with the promised sum, for which he had an order. Lord Ruthven's strength rapidly decreased; in two days mortification ensued, and death seemed advancing with hasty steps. His conduct and appearance had not changed; he seemed as unconscious of pain as he had been of the objects about him, but towards the close of the last evening, his mind became apparently uneasy, and his eye often fixed upon Aubrey, who was induced to offer his assistance with more than usual earnestness - "Assist me! You may save me - you may do more than that - I mean not life, I heed the death of my existence as little as that of the passing day; but you may save my honour, your friend's honour." - "How? Tell me how? I would do anything," replied Aubrey. "I need but little, my life ebbs apace - I cannot explain the whole - but if you would conceal all you know of me, my honour were free from stain in the world's mouth - and if my death were unknown for some time in England - I - I - but life." - "It shall not be known." - "Swear!" cried the dying man raising himself with exultant violence. "Swear by all your soul reveres, by all your nature fears, swear that for a year and a day you will not impart your knowledge of my crimes or death to any living being in any way, whatever may happen, or whatever you may see." His eyes seemed bursting from their sockets; "I swear!" said Aubrey; he sunk laughing upon his pillow, and breathed no more.

Aubrey retired to rest, but did not sleep; the many circumstances attending his acquaintance with this man rose upon his mind, and he knew not why; when he remembered his oath a cold shivering came over him, as if from the presentiment of something horrible awaiting him. Rising early in the morning, he was about to enter the hovel in which he had left the corpse, when a robber met him, and informed him that it was no longer there, having been conveyed by himself and comrades, upon his

Induce – *Persuade*
Ransom – *Payoff*
Conceal – *Hide*
Impart
– *Communicate*

retiring, to the pinnacle of a neighbouring mount, according to a promise they had given his lordship, that it should be exposed to the first cold ray of the moon that rose after his death. Aubrey astonished, and taking several of the men, determined to go and bury it upon the spot where it lay. But, when he had mounted to the summit he found no trace of either the corpse or the clothes, though the robbers swore they pointed out the identical rock on which they had laid the body. For a time his mind was bewildered in conjectures, but he at last returned, convinced that they had buried the corpse for the sake of the clothes.

Weary of a country in which he had met with such terrible misfortunes, and in which all apparently conspired to heighten that superstitious melancholy that had seized upon his mind, he resolved to leave it, and soon arrived at Smyrna. While waiting for a vessel to convey him to Otranto, or to Naples, he occupied himself in arranging those effects he had with him belonging to Lord Ruthven. Amongst other things there was a case containing several weapons of offence, more or less adapted to ensure the death of the victim. There were several daggers and ataghans. Whilst turning them over, and examining their curious forms, what was his surprise at finding a sheath apparently ornamented in the same style as the dagger discovered in the fatal hut. He shuddered; hastening to gain further proof, he found the weapon, and his horror may be imagined when he discovered that it fitted, though peculiarly shaped, the sheath he held in his hand. His eyes seemed to need no further certainty - they seemed gazing to be bound to the dagger, yet still he wished to disbelieve; but the particular form, the same varying tints upon the haft and sheath were alike in splendour on both, and left no room for doubt; there were also drops of blood on each. He left Smyrna, and on his way home, at Rome, his first inquiries were concerning the lady he had attempted to snatch from Lord Ruthven's seductive arts. Her parents were in distress, their fortune ruined, and she had not been heard of since the departure of his lordship. Aubrey's mind became almost broken under so many repeated horrors; he was afraid that this lady had fallen a victim to the destroyer of Ianthe. He became morose and silent; and his only occupation consisted in urging the speed of the postilions, as if he were going to save the life of someone he held dear. He arrived at Calais; a breeze, which seemed obedient to his will, soon wafted him to the English shores; and he hastened to the mansion of his fathers, and there, for a moment, appeared to lose, in the

Astonish – *Surprise*
Bewilder – *Puzzle*
Sheath – *Cover*

embraces and caresses of his sister, all memory of the past. If she before, by her infantine caresses, had gained his affection, now that the woman began to appear, she was still more attaching as a companion.

Miss Aubrey had not that winning grace which gains the gaze and applause of the drawing room assemblies. There was none of that light brilliancy which only exists in the heated atmosphere of a crowded apartment. Her blue eye was never lit up by the levity of the mind beneath. There was a melancholy charm about it which did not seem to arise from misfortune, but from some feeling within, that appeared to indicate a soul conscious of a brighter realm. Her step was not that light footing, which strays where a butterfly or a colour may attract - it was sedate and pensive. When alone, her face was never brightened by the smile of joy; but when her brother breathed to her his affection, and would in her presence forget those griefs she knew destroyed his rest, who would have exchanged her smile for that of the voluptuary? It seemed as if those eyes, that face were then playing in the light of their own native sphere. She was yet only eighteen, and had not been presented to the world, it having been thought by her guardians more fit that her presentation should be delayed until her brother's return from the continent, when he might be her protector. It was now, therefore, resolved that the next drawing room, which was fast approaching, should be the epoch of her entry into the "busy scene." Aubrey would rather have remained in the mansion of his fathers, and feed upon the melancholy which overpowered him. He could not feel interest about the frivolities of fashionable strangers, when his mind had been so torn by the events he had witnessed; but he determined to sacrifice his own comfort to the protection of his sister. They soon arrived in town, and prepared for the next day, which had been announced as a drawing room.

The crowd was excessive - a drawin groom had not been held for long time, and all who were anxious to bask in the smile of royalty, hastened thither. Aubrey was there with his sister. While he was standing in a corner by himself, heedless of all around him, engaged in the remembrance that the first time he had seen Lord Ruthven was in that very place - he felt himself suddenly seized by the arm, and a voice he recognized too well, sounded in his ear - "Remember your oath." He had hardly courage to turn, fearful of seeing a spectre that would blast him, when he perceived, at a little distance, the same figure which had attracted his notice on this spot upon his first entry into society. He gazed

Morose – *Miserable*
Sedate – *Tranquillize*
Pensive – *Thoughtful*
Frivolities –
Unsuitable behaviour

till his limbs almost refusing to bear their weight, he was obliged to take the arm of a friend, and forcing a passage through the crowd, he threw himself into his carriage, and was driven home. He paced the room with hurried steps, and fixed his hands upon his head, as if he were afraid his thoughts were bursting from his brain. Lord Ruthven again before him - circumstances started up in dreadful array - the dagger - his oath. He roused himself, he could not believe it possible - the dead rise again! He thought his imagination had conjured up the image his mind was resting upon. It was impossible that it could be real - he determined, therefore, to go again into society; for though he attempted to ask concerning Lord Ruthven, the name hung upon his lips and he could not succeed in gaining information. He went a few nights after with his sister to the assembly of a near relation. Leaving her under the protection of a matron, he retired into a recess, and there gave himself up to his own devouring thoughts. Perceiving, at last, that many were leaving, he roused himself, and entering another room, found his sister surrounded by several, apparently in earnest conversation; he attempted to pass and get near her, when one, whom he requested to move, turned round, and revealed to him those features he most abhorred. He sprang forward, seized his sister's arm, and, with hurried step, forced her towards the street. At the door he found himself impeded by the crowd of servants who were waiting for their lords; and while he was engaged in passing them, he again heard that voice whisper close to him - "Remember your oath!" - He did not dare to turn, but, hurrying his sister, soon reached home.

Aubrey became almost distracted. If before his mind had been absorbed by one subject, how much more completely was it engrossed now that the certainty of the monster's living again pressed upon his thoughts. His sister's attentions were now unheeded, and it was in vain that she intreated him to explain to her what had caused his abrupt conduct. He only uttered a few words, and those terrified her. The more he thought, the more he was bewildered. His oath startled him; was he then to allow this monster to roam, bearing ruin upon his breath, amidst all he held dear, and not avert its progress? His very sister might have been touched by him. But even if he were to break his oath, and disclose his suspicions, who would believe him? He thought of employing his own hand to free the world from such a wretch; but death, he remembered, had been already mocked. For days he remained in state; shut up in his room, he saw no one, and ate only when

Spectre – *Apparition*
Array – *Collection*
Conjure – *Juggle*
Abhor – *Detest*

his sister came, who, with eyes streaming with tears, besought him, for her sake, to support nature. At last, no longer capable of bearing stillness and solitude, he left his house, roamed from street to street, anxious to fly that image which haunted him. His dress became neglected, and he wandered, as often exposed to the noon-day sun as to the mid-night damps. He was no longer to be recognized; at first he returned with evening to the house; but at last he laid him down to rest wherever fatigue overtook him. His sister, anxious for his safety, employed people to follow him; but they were soon distanced by him who fled from a pursuer swifter than any from thought. His conduct, however, suddenly changed. Struck with the idea that he left by his absence the whole of his friends, with a fiend amongst them, of whose presence they were unconscious, he determined to enter again into society, and watch him closely, anxious to forewarn, in spite of his oath, all whom Lord Ruthven approached with intimacy. But when he entered into a room, his haggard and suspicious looks were so striking, his inward shuddering so visible, that his sister was at last obliged to beg of him to abstain from seeking, for her sake, a society which affected him so strongly. When, however, remonstrance proved unavailing, the guardians thought proper to interpose, and, fearing that his mind was becoming alienated, they thought it high time to resume again that trust which had been before imposed upon them by Aubrey's parents.

Desirous of saving him from the injuries and sufferings he had daily encountered in his wanderings, and of preventing him from exposing to the general eye those marks of what they considered folly, they engaged a physician to reside in the house, and take constant care of him. He hardly appeared to notice it, so completely was his mind absorbed by one terrible subject. His incoherence became at last so great that he was confined to his chamber. There he would often lie for days, incapable of being roused. He had become emaciated, his eyes had attained a glassy lustre; the only sign of affection and recollection remaining displayed itself upon the entry of his sister; then he would sometimes start, and, seizing her hands, with looks that severely afflicted her, he would desire her not to touch him. "Oh, do not touch him - if your love for me is aught, do not go near him!" When, however, she inquired to whom he referred, his only answer was, "True! true!" and again he sank into a state, whence not even she could rouse him. This lasted many months. Gradually, however, as the year was passing, his incoherences became less

Abrupt – *Sudden*
Besought – *Request*
Fatigue – *Exhaustion*
Abstain – *Refrain*

frequent, and his mind threw off a portion of its gloom, whilst his guardians observed, that several times in the day he would count upon his fingers a definite number, and then smile.

The time had nearly elapsed, when, upon the last day of the year, one of his guardians entering his room, began to converse with his physician upon the melancholy circumstance of Aubrey's being in so awful a situation, when his sister was going next day to be married. Instantly Aubrey's attention was attracted; he asked anxiously to whom. Glad of this mark of returning intellect, of which they feared he had been deprived, they mentioned the name of the Earl of Marsden. Thinking this was a young Earl whom he had met with in society, Aubrey seemed pleased, and astonished them still more by his expressing his intention to be present at the nuptials, and desiring to see his sister. They answered not, but in a few minutes his sister was with him. He was apparently again capable of being affected by the influence of her lovely smile; for he pressed her to his breast, and kissed her cheek, wet with tears, flowing at the thought of her brother's being once more alive to the feelings of affection. He began to speak with all his wonted warmth, and to congratulate her upon her marriage with a person so distinguished for rank and every accomplishment; when he suddenly perceived a locket upon her breast; opening it, what was his surprise at beholding the features of the monster who had so long influenced his life. He seized the portrait in a paroxysm of rage, and trampled it under foot. Upon her asking him why he thus destroyed the resemblance of her future husband, he looked as if he did not understand her; then seizing her hands, and gazing on her with a frantic expression of countenance, he bade her swear that she would never wed this monster, for he - But he could not advance - it seemed as if that voice again bade him remember his oath - he turned suddenly round, thinking Lord Ruthven was near him but saw no one. In the meantime the guardians and physician, who had heard the whole, and thought this was but a return of his disorder, entered, and forcing him from Miss Aubrey, desired her to leave him. He fell upon his knees to them, he implored, he begged of them to delay but for one day. They, attributing this to the insanity they imagined had taken possession of his mind endeavoured to pacify him, and retired.

Confine – *Detain*
Wonted
– *Customary*
Rage – *Anger*

Lord Ruthven had called the morning after the drawing room, and had been refused with everyone else. When he heard of Aubrey's ill health, he readily understood himself to be the cause of it; but when he learned that he was deemed insane, his exultation and pleasure could hardly be concealed from those among whom he had gained this information. He hastened to the house of his former companion, and, by constant attendance, and the pretence of great affection for the brother and interest in his fate, he gradually won the ear of Miss Aubrey. Who could resist his power? His tongue had dangers and toils to recount - could speak of himself as of an individual having no sympathy with any being on the crowded earth, save with her to whom he addressed himself; could tell how, since he knew her, his existence had begun to seem worthy of preservation, if it were merely that he might listen her soothing accents; in fact, he knew so well how to use the serpent's art, or such was the will of fate, that he gained her affections. The title of the elder branch falling at length to him, he obtained an important embassy, which served as an excuse for hastening the marriage (in spite of her brother's deranged state), which was to take place the very day before his departure for the continent.

Aubrey, when he left by the physician and his guardians, attempted to bribe the servants, but in vain. He asked for pen and paper; it was given him; he wrote a letter to his sister, conjuring her, as she valued her own happiness, her own honour, and the honour of those now in the grave, who once held her in their arms as their hope and the hope of their house, to delay but for a few hours that marriage, on which he denounced the most heavy curses. The servants promised they would deliver it; but giving it to the physician, he thought it better not to **harass** any more the mind of Miss Aubrey by, what he considered, the **ravings** of a maniac. Night passed on without rest to the busy inmates of the house; and Aubrey heard, with a horror that may more easily be conceived than described, the notes of busy preparation. Morning came, and the sound of carriages broke upon his ear. Aubrey grew almost **frantic**. The curiosity of the servants at last overcame their vigilance; they gradually stole away, leaving him in the custody of an helpless old woman. He seized the opportunity, with one bound was out of the room, and in a moment found himself in the apartment where all were nearly assembled. Lord Ruthven was the first to perceive him: he immediately approached, and, taking his arm by force, hurried him from the room, speechless with rage. When on the staircase, Lord Ruthven whispered in his ear - "Remember your oath, and know, if not my bride today, your sister is dishonoured. Women are frail!" So saying, he pushed him

Implore – *Beseech*
Pacify – *Appease*
Fate – *Fortune*
Derange – *Disturb*

towards his attendants, who, roused by the old woman, had come in search of him. Aubrey could no longer support himself; his rage not finding vent, had broken a blood vessel, and he was conveyed to bed. This was not mentioned to his sister, who was not present when he entered, as the physician was afraid of **agitating** her. The marriage was solemnized, and the bride and bridegroom left London.

Aubrey's weakness increased; the effusion of blood produced symptoms of the near approach of death. He desired his sister's guardians might be called, and when the midnight hour had struck, he related composedly what the reader has perused - he died immediately after. The guardians hastened to protect Miss Aubrey; but when they arrived, it was too late. Lord Ruthven had disappeared, and Aubrey's sister had glutted the thirst of a Vampyre!

Food For Thought

Who do you think was Ruthven actually? How did Aubrey's sister die? Who was responsible for her death? Do you think that Aubrey should have tried harder to save his sister? Can you suggest andy other ending to the story?

Harass – *Annoy*
Raving – *Wild*
Frantic – *Anxious*
Agitate – *Provoke*

An Understanding

Q. 1. Who is Aubrey and how does he meet Lord Ruthven? What sort of a man was Lord Ruthven and how was his character?

Ans. _____

_____

_____

_____

_____

Q. 2. What happens to Aubrey in Greece? Who is Ianthe and what did she tell Aubrey? What was the relationship between the two?

Ans. _____

_____

_____

_____

_____

Q. 3. How did Ianthe die in the story? Who do you think would have killed her?

Ans. _____

_____

_____

_____

_____

Q. 4. Aubrey and Ruthven are attacked by bandits. What happens next? What did Ruthven do before he died of mortal wounds?

Ans. _____

_____

_____

_____

_____

Edgar Allan Poe

Born on January 19, 1809
Died on October 7, 1849 (aged 40)
Notable Works:
Honours:

Early Life

Edgar Allan Poe was a well-known American author, poet, editor and literary critic, considered part of the American Romantic Movement. Poe was born in Boston, Massachusetts, where he was orphaned young when his mother died shortly after his father abandoned the family. Poe was taken in by John and Frances Allan, of Richmond, Virginia, but they never formally adopted him. He attended the University of Virginia for one semester, but left due to lack of money. After enlisting in the Army and later failing as an officer's cadet at West Point, Poe parted ways with the Allans. He married Virginia Clemm, his 13-year-old cousin in Baltimore in 1835.

Military Career

Unable to support himself, on May 27, 1827, Poe enlisted in the United States Army as a private. Using the name, "Edgar A. Perry", he claimed he was 22 years old, though he was 18. He first served at Fort Independence in Boston Harbor for five dollars a month.

Literary Works and Achievements

His publishing career began humbly, with an anonymous collection of poems, *Tamerlane and Other Poems (1827)*, credited only to "a Bostonian".

Poe switched his focus to prose and spent the next several years working for literary journals and periodicals, becoming known for his own style of literary criticism. In January 1845, Poe published his poem, "The Raven", to instant success. "The Bells" was also one of his famous poems. His wife died of tuberculosis, two years after its publication. He began planning to produce his own journal, *The Penn* (later renamed, *The Stylus*), though he died before it could be published.

Poe and his works influenced literature in the United States and around the world, as well as in specialised fields, such as cosmology and cryptography. Poe and his works appear throughout popular culture in literature, music, films and television. Some of his popular short stories are: "The Black Cat", "The Cask of Amontillado", The Gold-Bug", The "Hop-Frog", "The Masque of the Red Death", "The Oval Portrait", etc.

His other notable works are: *Politian* (1835) – Poe's only play, *The Narrative of Arthur Gordon, Pym of Nantucket (1838)* – Poe's only complete novel, "The Balloon-Hoax" (1844) – A journalistic hoax printed as a true story, "The Philosophy of Composition"

(1846) – Essay, "Eureka: A Prose Poem" (1848) – Essay, "The Poetic Principle" (1848) – Essay, etc.

Writing Style

Best known for his tales of mystery and the macabre, Poe was one of the earliest American practitioners of the short story and is considered the inventor of the *detective fiction genre*. He is further credited with contributing to the emerging *genre of science fiction*.

Later Works

Edgar Allan Poe's last incomplete work is "The Light-House" (1849). He died in Baltimore; the cause and circumstances that lead to his death remain certain. Edgar Allan Poe is buried in Baltimore, Maryland.

Trivia

A number of Edgar Allan Poe's homes are dedicated to museums today.

Manuscript Found in a Bottle
~ Edgar Allan Poe

Qui n'a plus qu'un moment a vivre
N'a plus rien a dissimuler
QUINAULT - *Atys*

OF my country and of my family I have little to say. Ill us-age and length of years have driven me from the one, and **estranged** me from the other. Hereditary wealth afford-ed me an education of no common order, and a contempla-tive turn of mind enabled me to methodize the stores which early study very diligently garnered up. Beyond all things, the study of the German moralists gave me great delight; not from any ill-advised admiration of their **eloquent** madness, but from the ease with which my habits of rigid thought en-abled me to detect their falsities. I have often been reproached with the aridity of my genius; a deficiency of imagination has been imputed to me as a crime; and the Pyrrhonism of my opinions has at all times rendered me notorious. Indeed, a strong relish for physical philosophy has, I fear, tinctured my mind with a very common error of this age - I mean the habit of referring occurrences, even the least susceptible of such reference, to the principles of that science. Upon the whole, no person could be less liable than myself to be led away from the severe **precincts** of truth by the ignes fatui of super-stition. I have thought proper to premise thus much, lest the incredible tale I have to tell should be considered rather the raving of a crude imagination, than the positive experience of a mind to which the reveries of fancy have been a dead letter and a nullity.

After many years spent in foreign travel, I sailed in the year 18--, from the port of Batavia, in the rich and populous island of Java, on a **voyage** to the Archipelago of the Sunda islands. I went as passenger - having no other inducement than a kind of nervous restlessness which haunted me as a fiend.

Our vessel was a beautiful ship of about four hundred tons, copper-fastened, and built at Bombay of Malabar teak. She was freighted with cotton-wool and oil, from the

Estrange – *Separate*
Eloquent
– *Expressive*
Precincts – *Confines*
Voyage – *Journey*

Lachadive islands. We had also on board coir, jaggeree, ghee, cocoa-nuts, and a few cases of opium. The storage was clumsily done, and the vessel consequently crank.

We got under way with a mere breath of wind, and for many days stood along the eastern coast of Java, without any other incident to **beguile** the monotony of our course than the occasional meeting with some of the small grabs of the Archipelago to which we were bound.

One evening, leaning over the taffrail, I observed a very singular, isolated cloud, to the N.W. It was remarkable, as well for its color, as from its being the first we had seen since our departure from Batavia. I watched it attentively until sunset, when it spread all at once to the eastward and westward, girting in the horizon with a narrow strip of vapor, and looking like a long line of low beach. My notice was soon afterwards attracted by the dusky-red appearance of the moon, and the peculiar character of the sea. The latter was undergoing a rapid change, and the water seemed more than usually transparent. Although I could distinctly see the bottom, yet, heaving the lead, I found the ship in fifteen **fathoms**. The air now became intolerably hot, and was loaded with spiral exhalations similar to those arising from heat iron. As night came on, every breath of wind died away, an more entire calm it is impossible to **conceive**. The flame of a candle burned upon the poop without the least perceptible motion, and a long hair, held between the finger and thumb, hung without the possibility of detecting a vibration. However, as the captain said he could perceive no indication of danger, and as we were drifting in bodily to shore, he ordered the sails to be furled, and the anchor let go. No watch was set, and the crew, consisting principally of Malays, stretched themselves deliberately upon deck. I went below - not without a full presentiment of evil. Indeed, every appearance warranted me in apprehending a Simoom. I told the captain my fears; but he paid no attention to what I said, and left me without deigning to give a reply. My uneasiness, however, prevented me from sleeping, and about midnight I went upon deck. As I placed my foot upon the upper step of the companion-ladder, I was startled by a loud, humming noise, like that occasioned by the rapid revolution of a mill-wheel, and before I could **ascertain** its meaning, I found the ship quivering to its centre. In the next instant, a wilderness of foam hurled us upon our beam ends,

Beguile – *Entice*
Fathom
– *Understand*
Conceive – *Consider*
Ascertain
– *Determine*

and, rushing over us fore and aft, swept the entire decks from stem to stern.

The extreme fury of the blast proved, in a great measure, the salvation of the ship. Although completely water-logged, yet, as her masts had gone by the board, she rose, after a minute, heavily from the sea, and, staggering awhile beneath the **immense** pressure of the tempest, finally righted.

By what miracle I escaped destruction, it is impossible to say. Stunned by the shock of the water, I found myself, upon recovery, jammed in between the stern-post and rudder. With great difficulty I gained my feet, and looking dizzily around, was, at first, struck with the idea of our being among breakers; so terrific, beyond the wildest imagination, was the whirlpool of mountainous and foaming ocean within which we were **engulfed**. After a while, I heard the voice of an old Swede, who had shipped with us at the moment of our leaving port. I hallooed to him with all my strength, and presently he came reeling aft. We soon discovered that we were the sole survivors of the accident. All on deck, with the exception of ourselves, had been swept overboard; the captain and mates must have **perished** as they slept, for the cabins were deluged with water. Without assistance, we could expect to do little for the security of the ship, and our exertions were at first paralyzed by the momentary expectation of going down. Our cable had, of course, parted like pack-thread, at the first breath of the hurricane, or we should have been instantaneously overwhelmed. We scudded with frightful velocity before the sea, and the water made clear breaches over us. The framework of our stern was shattered excessively, and, in almost every respect, we had received considerable injury; but to our extreme Joy we found the pumps unchoked, and that we had made no great shifting of our ballast. The main fury of the blast had already blown over, and we apprehended little danger from the violence of the wind; but we looked forward to its total cessation with **dismay**; well believing that in our shattered condition, we should inevitably perish in the tremendous swell which would ensue. But this very just apprehension seemed by no means likely to be soon verified. For five entire days and nights - during which our only subsistence was a small quantity of jaggeree, procured with great difficulty from the forecastle - the hulk flew at a rate defying

Immense – *Huge*
Engulf – *Overwhelm*
Perish – *Die*
Dismay – *Disappointment*

computation, before rapidly succeeding flaws of wind, which, without equalling the first violence of the Simoom, were still more terrific than any tempest I had before encountered. Our course for the first four days was, with trifling variations, S.E. and by S.; and we must have run down the coast of New Holland. On the fifth day the cold became extreme, although the wind had hauled round a point more to the northward. The sun arose with a sickly yellow **lustre**, and clambered a very few degrees above the horizon - emitting no decisive light. There were no clouds apparent, yet the wind was upon the increase, and blew with a fitful and unsteady fury. About noon, as nearly as we could guess, our attention was again arrested by the appearance of the sun. It gave out no light, properly so called, but a dull and **sullen** glow without reflection, as if all its rays were polarized. Just before sinking within the **turgid** sea, its central fires suddenly went out, as if hurriedly extinguished by some unaccountable power. It was a dim, sliver-like rim, alone, as it rushed down the unfathomable ocean.

We waited in vain for the arrival of the sixth day - that day to me has not arrived - to the Swede, never did arrive. Thenceforward we were enshrouded in patchy darkness, so that we could not have seen an object at twenty paces from the ship. **Eternal** night continued to envelop us, all unrelieved by the phosphoric sea-brilliancy to which we had been accustomed in the tropics. We observed too, that, although the tempest continued to rage with unabated violence, there was no longer to be discovered the usual appearance of surf, or foam, which had hitherto attended us. All around were horror, and thick gloom, and a black sweltering desert of ebony. Superstitious terror crept by degrees into the spirit of the old Swede, and my own soul was wrapped up in silent wonder. We neglected all care of the ship, as worse than useless, and securing ourselves, as well as possible, to the stump of the mizen-mast, looked out bitterly into the world of ocean. We had no means of calculating time, nor could we form any guess of our situation. We were, however, well aware of having made farther to the southward than any previous navigators, and felt great amazement at not meeting with the usual impediments of ice. In the meantime every moment threatened to be our last - every mountainous billow hurried to overwhelm us. The swell surpassed

Lustre – *Sheen*
Sullen – *Surly*
Turgid – *Pompous*
Eternal – *Everlasting*

anything I had imagined possible, and that we were not instantly buried is a miracle. My companion spoke of the lightness of our cargo, and reminded me of the excellent qualities of our ship; but I could not help feeling the utter hopelessness of hope itself, and prepared myself gloomily for that death which I thought nothing could defer beyond an hour, as with every knot of way the ship made, the swelling of the black stupendous seas became more dismally **appalling**. At times we gasped for breath at an elevation beyond the albatross, at times became dizzy with the velocity of our descent into some watery hell, where the air grew **stagnant**, and no sound disturbed the **slumbers** of the kraken.

We were at the bottom of one of these abysses, when a quick scream from my companion broke fearfully upon the night. "See! see!" cried he, shrieking in my ears, "Almighty God! See! See!" As he spoke, I became aware of a dull, sullen glare of red light which streamed down the sides of the vast chasm where we lay, and threw a fitful brilliancy upon our deck. Casting my eyes upwards, I beheld a spectacle which froze the current of my blood. At a terrific height directly above us, and upon the very verge of the precipitous descent, hovered a gigantic ship of, perhaps, four thousand tons. Although upreared upon the summit of a wave more than a hundred times her own altitude, her apparent size exceeded that of any ship of the line or East Indiaman in existence. Her huge hull was of a deep dingy black, unrelieved by any of the customary carvings of a ship. A single row of brass cannon protruded from her open ports, and dashed from their polished surfaces the fires of innumerable battle-lanterns, which swung to and fro about her rigging. But what mainly inspired us with horror and astonishment, was that she bore up under a press of sail in the very teeth of that supernatural sea, and of that ungovernable hurricane. When we first discovered her, her bows were alone to be seen, as she rose slowly from the dim and horrible gulf beyond her. For a moment of intense terror she paused upon the giddy pinnacle, as if in contemplation of her own sublimity, then trembled and tottered, and came down.

At this instant, I know not what sudden self-possession came over my spirit. Staggering as far aft as I could, I awaited fearlessly the ruin that was to **overwhelm**. Our own vessel was at length ceasing from her struggles, and sinking with her

Appalling – *Terrible*
Stagnant – *Sluggish*
Slumber – *Sleep*
Overwhelm – *Overpower*

head to the sea. The shock of the descending mass struck her, consequently, in that portion of her frame which was already under water, and the inevitable result was to **hurl** me, with irresistible violence, upon the rigging of the stranger.

As I fell, the ship hove in stays, and went about; and to the confusion ensuing I attributed my escape from the notice of the crew. With little difficulty I made my way unperceived to the main hatchway, which was partially open, and soon found an opportunity of secreting myself in the hold. Why I did so I can hardly tell. An indefinite sense of **awe**, which at first sight of the navigators of the ship had taken hold of my mind, was perhaps the principle of my concealment. I was unwilling to trust myself with a race of people who had offered, to the cursory glance I had taken, so many points of vague novelty, doubt, and apprehension. I therefore thought proper to contrive a hiding place in the hold. This I did by removing a small portion of the shifting boards, in such a manner as to afford me a convenient **retreat** between the huge timbers of the ship.

I had scarcely completed my work, when a footstep in the hold forced me to make use of it. A man passed by my place of concealment with a feeble and unsteady gait. I could not see his face, but had an opportunity of observing his general appearance. There was about it an evidence of great age and infirmity. His knees tottered beneath a load of years, and his entire frame **quivered** under the burthen. He muttered to himself, in a low broken tone, some words of a language which I could not understand, and groped in a corner among a pile of singular-looking instruments, and decayed charts of navigation. His manner was a wild mixture of the peevishness of second childhood, and the solemn dignity of a God. He at length went on deck, and I saw him no more.

A feeling, for which I have no name, had taken possession of my soul - a sensation which will admit of no analysis, to which the lessons of bygone times are inadequate, and for which I fear futurity itself will offer me no key. To a mind constituted like my own, the latter consideration is an evil. I shall never - I know that I shall never - be satisfied with regard to the nature of my conceptions. Yet it is not wonderful that these conceptions are indefinite, since they have their origin in sources so utterly novel. A new sense - a new entity is added to my soul.

Hurl – *Throw*
Awe – *Wonder*
Retreat – *Move away*
Quiver – *Tremble*

It is long since I first **trod** the deck of this terrible ship, and the rays of my destiny are, I think, gathering to a focus. **Incomprehensible** men! Wrapped up in meditations of a kind which I cannot divine, they pass me by unnoticed. Concealment is utter folly on my part, for the people will not see. It was but just now that I passed directly before the eyes of the mate - it was no long while ago that I ventured into the captain's own private cabin, and took thence the materials with which I write, and have written. I shall from time to time continue this journal. It is true that I may not find an opportunity of transmitting it to the world, but I will not fall to make the endeavour. At the last moment I will enclose the MS. in a bottle, and cast it within the sea.

An incident has occurred which has given me new room for meditation. Are such things the operation of ungoverned Chance? I had **ventured** upon deck and thrown myself down, without attracting any notice, among a pile of ratlin-stuff and old sails in the bottom of the yawl. While musing upon the singularity of my fate, I unwittingly daubed with a tar brush the edges of a neatly-folded studding sail which lay near me on a barrel. The studding sail is now bent upon the ship, and the thoughtless touches of the brush are spread out into the word DISCOVERY.

***

I have made many observations lately upon the structure of the vessel. Although well armed, she is not, I think, a ship of war. Her rigging, build, and general equipment, all negative a supposition of this kind. What she is not, I can easily perceive - what she is I fear it is impossible to say. I know not how it is, but in scrutinizing her strange model and singular cast of spars, her huge size and overgrown suits of canvas, her severely simple bow and antiquated stern, there will occasionally flash across my mind a sensation of familiar things, and there is always mixed up with such indistinct shadows of recollection, an unaccountable memory of old foreign chronicles and ages long ago.

***

I have been looking at the timbers of the ship. She is built of a material to which I am a stranger. There is a peculiar character

Trod – *Walked heavily*
Incomprehensible – *Impossible to understand*
Ventured – *Offered*

about the wood which strikes me as **rendering** it unfit for the purpose to which it has been applied. I mean its extreme porousness, considered independently by the worm-eaten condition which is a consequence of navigation in these seas, and apart from the rottenness attendant upon age. It will appear perhaps an observation somewhat over curious, but this wood would have every characteristic of Spanish oak, if Spanish oak were distended by any unnatural means.

In reading the above sentence a curious apothegm of an old weather-beaten Dutch navigator comes full upon my recollection. "It is as sure," he was wont to say, when any doubt was entertained of his veracity, "as sure as there is a sea where the ship itself will grow in bulk like the living body of the seaman."

***

About an hour ago, I made a bold move to **thrust** myself among a group of the crew. They paid me no manner of attention, and, although I stood in the very midst of them all, seemed utterly unconscious of my presence. Like the one I had at first seen in the hold, they all bore about them the marks of a hoary old age. Their knees trembled with infirmity; their shoulders were bent double with **decrepitude**; their shrivelled skins rattled in the wind; their voices were low, tremulous and broken; their eyes glistened with the rheum of years; and their gray hairs streamed terribly in the tempest. Around them, on every part of the deck, lay scattered mathematical instruments of the most quaint and obsolete construction.

***

I mentioned some time ago the bending of a studding sail. From that period the ship, being thrown dead off the wind, has continued her terrific course due south, with every rag of canvas packed upon her, from her trucks to her lower studding sail booms, and rolling every moment her top gallant yard arms into the most appalling hell of water which it can enter into the mind of a man to imagine. I have just left the deck, where I find it impossible to maintain a footing, although the crew seem to experience little inconvenience. It appears to me a miracle of miracles that our **enormous** bulk is not swallowed up at once and forever. We are surely doomed to hover continually upon

Render – *Reduce*
Thrust – *Shove*
Decrepitude – *Frailty*
Enormous – *Huge*

the brink of Eternity, without taking a final **plunge** into the abyss. From billows a thousand times more stupendous than any I have ever seen, we glide away with the facility of the arrowy sea gull; and the colossal waters rear their heads above us like demons of the deep, but like demons **confined** to simple threats and forbidden to destroy. I am led to attribute these frequent escapes to the only natural cause which can account for such effect. I must suppose the ship to be within the influence of some strong current, or impetuous under-tow.

***

I have seen the captain face to face, and in his own cabin - but, as I expected, he paid me no attention. Although in his appearance there is, to a casual observer, nothing which might **bespeak** him more or less than man — still a feeling of irrepressible reverence and awe mingled with the sensation of wonder with which I regarded him. In stature he is nearly my own height; that is, about five feet eight inches. He is of a well-knit and compact frame of body, neither robust nor remarkably otherwise. But it is the singularity of the expression which reigns upon the face - it is the intense, the wonderful, the thrilling evidence of old age, so utter, so extreme, which excites within my spirit a sense - a sentiment ineffable. His forehead, although little wrinkled, seems to bear upon it the stamp of a myriad of years. His gray hairs are records of the past, and his grayer eyes are Sybils of the future. The cabin floor was thickly **strewn** with strange, iron-clasped folios, and mouldering instruments of science, and **obsolete** long-forgotten charts. His head was bowed down upon his hands, and he pored, with a fiery unquiet eye, over a paper which I took to be a commission, and which, at all events, bore the signature of a monarch. He muttered to himself, as did the first seaman whom I saw in the hold, some low peevish syllables of a foreign tongue, and although the speaker was close at my elbow, his voice seemed to reach my ears from the distance of a mile.

***

The ship and all in it are imbued with the spirit of Eld. The crew glide to and fro like the ghosts of buried centuries; their eyes have an eager and uneasy meaning; and when their

Plunge – *Dive*
Confine – *Limit*
Bespeak – *Signify*
Strewn – *Scattered*
Obsolete – *Outdated*

fingers fall **athwart** my path in the wild glare of the battle lanterns, I feel as I have never felt before, although I have been all my life a dealer in antiquities, and have imbibed the shadows of fallen columns at Balbec, and Tadmor, and Persepolis, until my very soul has become a ruin.

***

When I look around me I feel ashamed of my former apprehensions. If I trembled at the blast which has hitherto attended us, shall I not stand aghast at a warring of wind and ocean, to convey any idea of which the words tornado and simoom are trivial and ineffective? All in the immediate vicinity of the ship is the blackness of eternal night, and a chaos of foamless water; but, about a league on either side of us, may be seen, indistinctly and at intervals, **stupendous** ramparts of ice, towering away into the desolate sky, and looking like the walls of the universe.

***

As I imagined, the ship proves to be in a current; if that appellation can properly be given to a tide which, howling and shrieking by the white ice, thunders on to the southward with a velocity like the headlong dashing of a cataract.

To **conceive** the horror of my sensations is, I presume, utterly impossible; yet a curiosity to penetrate the mysteries of these awful regions, predominates even over my despair, and will reconcile me to the most hideous aspect of death. It is evident that we are hurrying onwards to some exciting knowledge - some never-to-be-imparted secret, whose attainment is destruction. Perhaps this current leads us to the southern pole itself. It must be confessed that a supposition apparently so wild has every probability in its favor.

***

Athwart – *Crossways*
Stupendous –
Astonishing
Conceive – *Consider*
Tremulous –
Timorous
Apathy – *Indifference*

The crew pace the deck with unquiet and **tremulous** step; but there is upon their countenances an expression more of the eagerness of hope than of the **apathy** of despair.

In the meantime the wind is still in our poop, and, as we carry a crowd of canvas, the ship is at times lifted bodily from out the sea - Oh, horror upon horror! the ice opens suddenly to the right, and to the left, and we are whirling dizzily, in immense concentric circles, round and round the borders of a gigantic amphitheatre, the summit of whose walls is lost in the darkness and the distance. But little time will be left me to **ponder** upon my destiny - the circles rapidly grow small - we are plunging madly within the grasp of the whirlpool - and amid a roaring, and bellowing, and thundering of ocean and of tempest, the ship is quivering, oh God! and - going down.

Note. - The "MS. Found in a Bottle," was originally published in 1831, and it was not until many years afterwards that I became **acquainted** with the maps of Mercator, in which the ocean is represented as rushing, by four mouths, into the (northern) Polar Gulf, to be absorbed into the bowels of the earth; the Pole itself being represented by a black rock, towering to a **prodigious** height. [Poe]

Food For Thought

The cargo ship from Batavia presently known as Jakarta in Indonesia is hit by a Simoon? What do you think a Simoon is? Do you believe that when a simoon hits a ship, anybody can survive? Give reasons for your answer.

Ponder – *Think deeply*
Acquaint – *Familiarise*
Prodigious – *Extraordinary*

An Understanding

Q. 1. Why has the story been named as 'Manuscript found in a Bottle' by the author? What is the plot of the story in brief?

Ans. _____

_____

_____

_____

_____

Q. 2. From where does the unnamed narrator of the story start his journey? By which means of transport does he travel through his journey and what happens to him?

Ans. _____

_____

_____

_____

_____

Q. 3. Where was the ship being driven to? What happens when the ship is about to reach Antarctica?

Ans. _____

_____

_____

_____

_____

Q. 4. How does the narrator manage to pen down his adventures and harrowing experiences befor his own disastrous death?

Ans. _____

_____

_____

_____

_____

Jack London

Born on January 12, 1876
Died on November 22, 1916 (aged 40)
Genres: Realism and Naturalism
Notable Works: London's most famous novels are *The Call of the Wild*, *White Fang*, *The Sea- Wolf*, *The Iron Heel*, and *Martin Eden*.

Early Life

John Griffith "Jack" London was born as John Griffith Chaney on January 12, 1876. He was an American author, journalist, and social activist. He was also a pioneer in the then-burgeoning world of commercial magazine fiction and was one of the *first fiction writers* to obtain worldwide celebrity and a large fortune from his fiction alone. He is best remembered as the author of the *Call of the Wild* and *White Fang*, both set in the Klondike Gold Rush, as well as the short stories "To Build a Fire", "An Odyssey of the North", and "Love of Life". He was the fifth and youngest child of Pennsylvania Canal builder Marshall Wellman and his first wife, Eleanor Garrett Jones. Jack London's house burnt down in the fire after the 1906 San Francisco earthquake; the California Historical Society placed a plaque at the site in 1953. Though the family was working class, it was not as impoverished as London's later accounts claimed. London was essentially self-educated.

In 1885 London found and read Ouida's long Victorian novel, *Signa*. He credited this as the seed of his literary success. In 1889, London began working 12 to 18 hours a day at Hickmott's Cannery. Seeking a way out, he borrowed money from his black foster mother Virginia Prentiss, bought the sloop Razzle-Dazzle from an oyster pirate named French Frank, and became an oyster pirate. In 1893, he signed on to the sealing schooner Sophie Sutherland, bound for the coast of Japan. When he returned, the country was in the grip of the panic of '93 and Oakland was swept by labor unrest. After grueling jobs in a jute mill and a street-railway power plant, he joined Kelly's Army and began his career as a tramp.

After many experiences as a hobo and a sailor, he returned to Oakland and attended the Oakland High School. He contributed a number of articles to the high school's magazine, *The Aegis*. His first published work was *Typhoon off the Coast of Japan*, an account of his sailing experiences.

London desperately wanted to attend the University of California, Berkeley. In 1896, after a summer of intense studying to pass certification exams, he was admitted. Financial circumstances forced him to leave in 1897 and he never graduated.

Literary Works and Achievements

London was fortunate in the timing of his writing career. He started just as new printing technologies enabled lower-cost production of magazines. This resulted in a boom in popular magazines aimed at a wide public, and a strong market for short fiction. In 1900, he made $2,500 in writing, about $70,000 in current value. His career was well under way.

Among the works, he sold to magazines was a short story known as either "Batard" or "Diable", in two editions of the same basic story.

London was a passionate advocate of unionisation, socialism, and the rights of workers and wrote several powerful works dealing with these topics such as his dystopian novel, *The Iron Heel* and his non-fiction exposé, *The People of the Abyss*. He also wrote of the South Pacific in such stories as *The Pearls of Parlay* and *The Heathen*, and of the San Francisco Bay area in *The Sea Wolf*.

Later Years

In later life, Jack London indulged his wide-ranging interests by accumulating a personal library of 15,000 volumes. He referred to his books as "the tools of my trade."

Many sources describe London's death as a suicide. This conjecture appears to be a rumor, or speculation based on incidents in his fiction writings. His death certificate gives the cause as uremia, following acute renal colic, a type of pain commonly caused by kidney stones. Uremia is also known as uremic poisoning.

Trivia

Jack London married twice and had two daughters, Joan and Becky by his first wife, Bessie Maddern London.

The Enemy of All the World

~ Jack London

IT was Silas Bannerman who finally ran down that scientific wizard and arch-enemy of mankind, Emil Gluck. Gluck's **confession**, before he went to the electric chair, threw much light upon the series of mysterious events, many apparently unrelated, that so **perturbed** the world between the years 1933 and 1941. It was not until that remarkable document was made public that the world dreamed of there being any connection between the assassination of the King and Queen of Portugal and the murders of the New York City police officers. While the deeds of Emil Gluck were all that was abominable, we cannot but feel, to a certain extent, pity for the unfortunate, malformed, and maltreated genius. This side of his story has never been told before, and from his confession and from the great mass of evidence and the documents and records of the time we are able to construct a fairly accurate portrait of him, and to **discern** the factors and pressures that moulded him into the human monster he became and that drove him onward and downward along the fearful path he trod. Emil Gluck was born in Syracuse, New York, in 1895. His father, Josephus Gluck, was a special policeman and night watchman, who, in the year 1900, died suddenly of pneumonia. The mother, a pretty, fragile creature, who, before her marriage, had been a milliner, **grieved** herself to death over the loss of her husband. This sensitiveness of the mother was the heritage that in the boy became morbid and horrible.

In 1901, the boy, Emil, then six years of age, went to live with his aunt, Mrs. Ann Bartell. She was his mother's sister, but in her breast was no kindly feeling for the sensitive, shrinking boy. Ann Bartell was a vain, shallow, and heartless woman. Also, she was cursed with poverty and burdened with a husband who was a lazy, erratic ne'er-do-well. Young Emil Gluck was not wanted, and Ann Bartell could be trusted to impress this fact sufficiently upon him. As an illustration of the treatment he received in that early, formative period, the following instance is given.

When he had been living in the Bartell home a little more than a year, he broke his leg. He sustained the injury

Confession –
Revelation
Perturb – *Trouble*
Discern – *Separate*
Grieve – *Mourn*

through playing on the forbidden roof - as all boys have done and will continue to do to the end of time. The leg was broken in two places between the knee and thigh. Emil, helped by his frightened playmates, managed to drag himself to the front sidewalk, where he fainted. The children of the neighbourhood were afraid of the hard-featured shrew who presided over the Bartell house; but, summoning their **resolution**, they rang the bell and told Ann Bartell of the accident. She did not even look at the little lad who lay stricken on the sidewalk, but slammed the door and went back to her wash- tub. The time passed. A drizzle came on, and Emil Gluck, out of his faint, lay sobbing in the rain. The leg should have been set immediately. As it was, the inflammation rose rapidly and made a nasty case of it. At the end of two hours, the **indignant** women of the neighbourhood protested to Ann Bartell. This time she came out and looked at the lad. Also she kicked him in the side as he lay helpless at her feet, and she hysterically disowned him. He was not her child, she said, and recommended that the ambulance be called to take him to the city receiving hospital. Then she went back into the house.

It was a woman, Elizabeth Shepstone, who came along, learned the situation, and had the boy placed on a shutter. It was she who called the doctor, and who, brushing aside Ann Bartell, had the boy carried into the house. When the doctor arrived, Ann Bartell promptly warned him that she would not pay him for his services. For two months the little Emil lay in bed, the first month on his back without once being turned over; and he lay neglected and alone, save for the occasional visits of the unremunerated and over-worked physician. He had no toys, nothing with which to **beguile** the long and **tedious** hours. No kind word was spoken to him, no soothing hand laid upon his brow, no single touch or act of loving tenderness - naught but the reproaches and harshness of Ann Bartell, and the continually reiterated information that he was not wanted. And it can well be understood, in such environment, how there was generated in the lonely, neglected boy much of the bitterness and hostility for his kind that later was to express itself in deeds so frightful as to terrify the world.

It would seem strange that, from the hands of Ann Bartell, Emil Gluck should have received a college education; but the explanation is simple. Her ne'er-do-well husband, deserting

Resolution – *Resolve*
Indignant – *Angry*
Beguile – *Entice*
Tedious – *Boring, tiresome*

her, made a strike in the Nevada goldfields, and returned to her a many-times millionaire. Ann Bartell hated the boy, and immediately she sent him to the Farristown Academy, a hundred miles away. Shy and sensitive, a lonely and misunderstood little soul, he was more lonely than ever at Farristown. He never came home, at vacation, and holidays, as the other boys did. Instead, he **wandered** about the deserted buildings and grounds, befriended and misunderstood by the servants and gardeners, reading much, it is remembered, spending his days in the fields or before the fire place with his nose poked always in the pages of some book. It was at this time that he over used his eyes and was compelled to take up the wearing of glasses, which same were so prominent in the photographs of him published in the newspapers in 1941.

He was a remarkable student. Application such as his would have taken him far; but he did not need application. A glance at a text meant mastery for him. The result was that he did an immense amount of collateral reading and acquired more in half a year than did the average student in half-a-dozen years. In 1909, barely fourteen years of age, he was ready - "more than ready" the headmaster of the academy said - to enter Yale or Harvard. His **juvenility** prevented him from entering those universities, and so, in 1909, we find him a freshman at historic Bowdoin College. In 1913 he graduated with highest honours, and immediately afterward followed Professor Bradlough to Berkeley, California. The one friend that Emil Gluck discovered in all his life was Professor Bradlough. The latter's weak lungs had led him to exchange Maine for California, the removal being facilitated by the offer of a professorship in the State University. Throughout the year 1914, Emil Gluck resided in Berkeley and took special scientific courses. Towards the end of that year two deaths changed his **prospects** and his relations with life. The death of Professor Bradlough took from him the one friend he was ever to know, and the death of Ann Bartell left him penniless. Hating the unfortunate lad to the last, she cut him off with one hundred dollars.

The following year, at twenty years of age, Emil Gluck was enrolled as an instructor of chemistry in the University of California. Here the years passed quietly; he faithfully performed the **drudgery** that brought him his salary, and, a student always, he took half-a-dozen degrees. He was, among

Wander – *Stroll*
Juvenile – *Young*
Prospect – *View*
Drudgery – *Labour*

other things, a Doctor of Sociology, of Philosophy, and of Science, though he was known to the world, in later days, only as Professor Gluck.

He was twenty-seven years old when he first sprang into prominence in the newspapers through the publication of his book, Sex and Progress. The book remains today a milestone in the history and philosophy of marriage. It is a heavy tome of over seven hundred pages, painfully careful and accurate, and startlingly original. It was a book for scientists, and not one calculated to make a stir. But Gluck, in the last chapter, using barely three lines for it, mentioned the hypothetical desirability of trial marriages. At once the newspapers seized these three lines, "played them up yellow," as the slang was in those days, and set the whole world laughing at Emil Gluck, the bespectacled young professor of twenty-seven. Photographers snapped him, he was besieged by reporters, women's clubs throughout the land passed resolutions condemning him and his immoral theories; and on the floor of the California Assembly, while discussing the state appropriation to the University, a motion demanding the expulsion of Gluck was made under threat of withholding the appropriation - of course, none of his **persecutors** had read the book; the twisted newspaper version of only three lines of it was enough for them. Here began Emil Gluck's hatred for newspaper men. By them his serious and intrinsically valuable work of six years had been made a laughing-stock and a **notoriety**. To his dying day, and to their everlasting **regret**, he never forgave them.

It was the newspapers that were responsible for the next disaster that befell him. For the five years following the publication of his book he had remained silent, and silence for a lonely man is not good. One can **conjecture** sympathetically the awful solitude of Emil Gluck in that populous University; for he was without friends and without sympathy. His only recourse was books, and he went on reading and studying enormously. But in 1927 he accepted an invitation to appear before the Human Interest Society of Emeryville. He did not trust himself to speak, and as we write we have before us a copy of his learned paper. It is sober, scholarly, and scientific, and, it must also be added, conservative. But in one place he dealt with, and I quote his words, "the industrial and social revolution that is taking place in society." A reporter present seized upon the word "revolution," divorced it from the text,

Persecutor – *Tyrant*
Notoriety – *Infamy*
Regret – *Remorse*
Conjecture – *Guesswork*

and wrote a **garbled** account that made Emil Gluck appear an anarchist. At once, "Professor Gluck, anarchist," flamed over the wires and was appropriately "featured" in all the newspapers in the land.

He had attempted to reply to the previous newspaper attack, but now he remained silent. Bitterness had already **corroded** his soul. The University faculty appealed to him to defend himself, but he sullenly declined, even refusing to enter in defence a copy of his paper to save himself from expulsion. He refused to resign, and was discharged from the University faculty. It must be added that political pressure had been put upon the University Regents and the President.

Persecuted, maligned, and misunderstood, the **forlorn** and lonely man made no attempt at **retaliation**. All his life he had been sinned against, and all his life he had sinned against no one. But his cup of bitterness was not yet full to overflowing. Having lost his position, and being without any income, he had to find work. His first place was at the Union Iron Works, in San Francisco, where he proved a most able draughtsman. It was here that he obtained his firsthand knowledge of battleships and their construction. But the reporters discovered him and featured him in his new vocation. He immediately resigned and found another place; but after the reporters had driven him away from half-a-dozen positions, he steeled himself to brazen out the newspaper persecution. This occurred when he started his electroplating establishment - in Oakland, on Telegraph Avenue. It was a small shop, employing three men and two boys. Gluck himself worked long hours. Night after night, as Policeman Carew testified on the stand, he did not leave the shop till one and two in the morning. It was during this period that he perfected the improved ignition device for gas-engines, the royalties from which ultimately made him wealthy.

He started his electroplating establishment early in the spring of 1928, and it was in the same year that he formed the disastrous love attachment for Irene Tackley. Now it is not to be imagined that an extraordinary creature such as Emil Gluck could be any other than an extraordinary lover. In addition to his genius, his loneliness, and his morbidness, it must be taken into consideration that he knew nothing about women. Whatever tides of desire flooded his being, he was unschooled in the conventional expression of them; while his excessive

Garble – *Jumble*
Corrode – *Rust*
Forlorn – *Lonely*
Retaliate – *React*

timidity was bound to make his love-making unusual. Irene Tackley was a rather pretty young woman, but shallow and light-headed. At the time she worked in a small candy store across the street from Gluck's shop. He used to come in and drink ice-cream sodas and lemon squashes, and stare at her. It seems the girl did not care for him, and merely played with him. He was "queer," she said; and at another time she called him a crank when describing how he sat at the counter and peered at her through his spectacles, blushing and stammering when she took notice of him, and often leaving the shop in **precipitate** confusion.

Gluck made her the most amazing presents - a silver tea-service, a diamond ring, a set of furs, opera glasses, a ponderous History of the World in many volumes, and a motor-cycle all silver-plated in his own shop. Enters now the girl's lover, putting his foot down, showing great anger, compelling her to return Gluck's strange assortment of presents. This man, William Sherbourne, was a gross and stolid creature, a heavy-jawed man of the working class who had become a successful building-contractor in a small way. Gluck did not understand. He tried to get an explanation, attempting to speak with the girl when she went home from work in the evening. She complained to Sherbourne, and one night he gave Gluck a beating. It was a very severe beating, for it is on the records of the Red Cross Emergency Hospital that Gluck was treated there that night and was unable to leave the hospital for a week.

Still Gluck did not understand. He continued to **seek** an explanation from the girl. In fear of Sherbourne, he applied to the Chief of Police for permission to carry a revolver, which permission was refused, the newspapers as usual playing it up **sensationally**. Then came the murder of Irene Tackley, six days before her contemplated marriage with Sherbourne. It was on a Saturday night. She had worked late in the candy store, departing after eleven o'clock with her week's wages in her purse. She rode on a San Pablo Avenue surface car to Thirty-fourth Street, where she alighted and started to walk the three blocks to her home. That was the last seen of her alive. Next morning she was found, strangled, in a vacant lot.

Emil Gluck was immediately arrested. Nothing that he could do could save him. He was convicted, not merely on circumstantial evidence, but on evidence "cooked up" by the

Precipitate – *Hurried*
Seek – *Pursue*
Sensational – *Extraordinary*

Oakland police. There is no discussion but that a large portion of the evidence was manufactured. The testimony of Captain Shehan was the sheerest **perjury**, it being proved long afterward that on the night in question he had not only not been in the vicinity of the murder, but that he had been out of the city in a resort on the San Leandro Road. The unfortunate Gluck received life imprisonment in San Quentin, while the newspapers and the public held that it was a miscarriage of justice - that the death penalty should have been visited upon him.

Gluck entered San Quentin prison on April 17, 1929. He was then thirty-four years of age. And for three years and a half, much of the time in solitary confinement, he was left to meditate upon the injustice of man. It was during that period that his bitterness corroded home and he became a hater of all his kind. Three other things he did during the same period: he wrote his famous treatise, Human Morals, his remarkable brochure, The Criminal Sane, and he worked out his awful and monstrous scheme of revenge. It was an episode that had occurred in his electroplating establishment that suggested to him his unique weapon of revenge. As stated in his confession, he worked every detail out theoretically during his imprisonment, and was able, on his release, immediately to embark on his career of **vengeance**.

His release was sensational. Also it was miserably and criminally delayed by the soulless legal red tape then in vogue. On the night of February 1, 1932, Tim Haswell, a hold-up man, was shot during an attempted robbery by a citizen of Piedmont Heights. Tim Haswell lingered three days, during which time he not only confessed to the murder of Irene Tackley, but furnished conclusive proofs of the same. Bert Danniker, a convict dying of consumption in Folsom Prison, was **implicated** as accessory, and his confession followed. It is inconceivable to us of today - the bungling, **dilatory** processes of justice a generation ago. Emil Gluck was proved in February to be an innocent man, yet he was not released until the following October. For eight months, a greatly wronged man, he was compelled to undergo his **unmerited** punishment. This was not conducive to sweetness and light, and we can well imagine how he ate his soul with bitterness during those dreary eight months.

He came back to the world in the fall of 1932, as usual a "feature" topic in all the newspapers. The papers, instead

Perjury –
Untruthfulness
Vengeance
– Retaliation, Revenge
Implicate – *Involve*
Dilatory – *Slow*
Unmerited –
Undeserved

of expressing heartfelt regret, continued their old sensa-tional persecution. One paper did more - the San Francisco Intelligencer. John Hartwell, its editor, elaborated an **ingenious** theory that got around the confessions of the two criminals and went to show that Gluck was responsible, after all, for the murder of Irene Tackley. Hartwell died. And Sherbourne died too, while Policeman Phillipps was shot in the leg and discharged from the Oakland police force.

The murder of Hartwell was long a mystery. He was alone in his editorial office at the time. The reports of the revolver were heard by the office boy, who rushed in to find Hartwell expiring in his chair. What puzzled the police was the fact, not merely that he had been shot with his own revolver, but that the revolver had been exploded in the drawer of his desk. The bullets had torn through the front of the drawer and entered his body. The police **scouted** the theory of suicide, murder was dismissed as **absurd**, and the blame was thrown upon the Eureka Smokeless Cartridge Company. Spontaneous explosion was the police explanation, and the chemists of the cartridge company were well bullied at the inquest. But what the police did not know was that across the street, in the Mercer Building, Room 633, rented by Emil Gluck, had been occupied by Emil Gluck at the very moment Hartwell's revolver so mysteriously exploded.

At the time, no connection was made between Hartwell's death and the death of William Sherbourne. Sherbourne had continued to live in the home he had built for Irene Tackley, and one morning in January, 1933, he was found dead. Suicide was the **verdict** of the coroner's **inquest**, for he had been shot by his own revolver. The curious thing that happened that night was the shooting of Policeman Phillipps on the sidewalk in front of Sherbourne's house. The policeman crawled to a police telephone on the corner and rang up for an ambulance. He claimed that someone had shot him from behind in the leg. The leg in question was so badly shattered by three '38 calibre bullets that amputation was necessary. But when the police discovered that the damage had been done by his own revol-ver, a great laugh went up, and he was charged with having been drunk. In spite of his denial of having touched a drop, and of his persistent assertion that the revolver had been in his hip pocket and that he had not laid a finger to it, he was discharged from the force. Emil Gluck's confession, six years

Ingenious – *Clever*
Scout – *Lookout*
Absurd – *Ridiculous*
Inquest – *Investigation*
Verdict – *Decision*

Greatest Science Fiction Stories

later, cleared the unfortunate policeman of disgrace, and he is alive today and in good health, the **recipient** of a handsome pension from the city.

Emil Gluck, having disposed of his immediate enemies, now sought a wider field, though his **enmity** for newspaper men and for the police remained always active. The royalties on his ignition device for gasoline engines had mounted up while he lay in prison, and year by year the earning power of his invention increased. He was independent, able to travel wherever he willed over the earth and to glut his monstrous appetite for revenge. He had become a monomaniac and an anarchist - not a philosophic anarchist, merely, but a violent anarchist. Perhaps the word is misused, and he is better described as a nihilist, or an annihilist. It is known that he affiliated with none of the groups of terrorists. He operated wholly alone, but he created a thousandfold more terror and achieved a thousandfold more destruction than all the terrorist groups added together.

He signalised his departure from California by blowing up Fort Mason. In his confession he spoke of it as a little experiment - he was merely trying his hand. For eight years he wandered over the earth, a mysterious terror, destroying property to the tune of hundreds of millions of dollars, and destroying countless lives. One good result of his awful **deeds** was the destruction he wrought among the terrorists themselves. Every time he did anything the terrorists in the vicinity were gathered in by the police **dragnet**, and many of them were executed. Seventeen were executed at Rome alone, following the assassination of the Italian King.

Perhaps the most world-amazing achievement of his was the assassination of the King and Queen of Portugal. It was their wedding day. All possible precautions had been taken against the terrorists, and the way from the cathedral, through Lisbon's streets, was double-banked with troops, while a squad of two hundred mounted troopers surrounded the carriage. Suddenly the amazing thing happened. The automatic rifles of the troopers began to go off, as well as the rifles, in the immediate vicinity, of the double-banked infantry. In the excitement the muzzles of the exploding rifles were turned in all directions. The slaughter was terrible - horses, troops, spectators, and the King and Queen, were riddled with bullets. To complicate the affair, in different

Recipient – *Receiver*
Enmity – *Hostility, Hatred*
Deed – *Action*
Dragnet – *Pursuit*

parts of the crowd behind the foot-soldiers, two terrorists had bombs explode on their persons. These bombs they had intended to throw if they got the opportunity. But who was to know this? The frightful havoc wrought by the bursting bombs but added to the confusion; it was considered part of the general attack.

One puzzling thing that could not be explained away was the conduct of the troopers with their exploding rifles. It seemed impossible that they should be in the plot, yet there were the hundreds their flying bullets had slain, including the King and Queen. On the other hand, more baffling than ever was the fact that seventy per cent. of the troopers themselves had been killed or wounded. Some explained this on the ground that the loyal foot-soldiers, witnessing the attack on the royal carriage, had opened fire on the **traitors**. Yet not one bit of evidence to verify this could be drawn from the survivors, though many were put to the torture. They contended **stubbornly** that they had not discharged their rifles at all, but that their rifles had discharged themselves. They were laughed at by the chemists, who held that, while it was just barely probable that a single cartridge, charged with the new smokeless powder, might spontaneously explode, it was beyond all probability and possibility for all the cartridges in a given area, so charged, **spontaneously** to explode. And so, in the end, no explanation of the amazing occurrence was reached. The general opinion of the rest of the world was that the whole affair was a blind panic of the feverish Latins, precipitated, it was true, by the bursting of two terrorist bombs; and in this connection was recalled the laughable **encounter** of long years before between the Russian fleet and the English fishing boats.

And Emil Gluck chuckled and went his way. He knew. But how was the world to know? He had stumbled upon the secret in his old electroplating shop on Telegraph Avenue in the city of Oakland. It happened, at that time, that a wireless telegraph station was established by the Thurston Power Company close to his shop. In a short time his electroplating vat was put out of order. The vat-wiring had many bad joints, and, on investigation, Gluck discovered minute welds at the joints in the wiring. These, by lowering the resistance, had caused an excessive current to pass through the solution, "boiling" it and spoiling the work. But what had caused the

Traitor – *Conspirator*
Stubborn – *Persistent*
Spontaneous – *Impulsive*
Encounter – *Meeting*

welds? was the question in Gluck's mind. His reasoning was simple. Before the establishment of the wireless station, the vat had worked well. Not until after the establishment of the wireless station had the vat been ruined. Therefore the wireless station had been the cause. But how? He quickly answered the question. If an electric discharge was capable of operating a coherer across three thousand miles of ocean, then, certainly, the electric discharges from the wireless station four hundred feet away could produce coherer effects on the bad joints in the vat wiring.

Gluck thought no more about it at the time. He merely re-wired his vat and went on electroplating. But afterwards, in prison, he remembered the incident, and like a flash there came into his mind the full significance of it. He saw in it the silent, secret weapon with which to revenge himself on the world. His great discovery, which died with him, was control over the direction and scope of the electric discharge. At the time, this was the unsolved problem of wireless telegraphy - as it still is today - but Emil Gluck, in his prison cell, mastered it. And, when he was released, he applied it. It was fairly simple, given the directing power that was his, to introduce a spark into the powder magazines of a fort, a battleship, or a revolver. And not alone could he thus explode powder at a distance, but he could ignite conflagrations. The great Boston fire was started by him - quite by accident, however, as he stated in his confession, adding that it was a pleasing accident and that he had never had any reason to regret it.

It was Emil Gluck that caused the terrible German-American War, with the loss of 800,000 lives and the consumption of almost **incalculable** treasure. It will be remembered that in 1939, because of the Pickard incident, strained relations existed between the two countries. Germany, though **aggrieved**, was not **anxious** for war, and, as a peace token, sent the Crown Prince and seven battleships on a friendly visit to the United States. On the night of February 15, the seven warships lay at anchor in the Hudson opposite New York City. And on that night Emil Gluck, alone, with all his apparatus on board, was out in a launch. This launch, it was afterwards proved, was bought by him from the Ross Turner Company, while much of the **apparatus** he used that night had been purchased from the Columbia Electric Works. But this was not known at the time.

Incalculable –
Innumerable
Aggrieved – *Hurt*
Anxious – *Nervous*
Apparatus – *Device*

All that was known was that the seven battleships blew up, one after another, at regular four-minute intervals. Ninety per cent of the **crews** and officers, along with the Crown Prince, perished. Many years before, the American battleship Maine had been blown up in the harbour of Havana, and war with Spain had immediately followed - though there has always existed a reasonable doubt as to whether the explosion was due to **conspiracy** or accident. But accident could not explain the blowing up of the seven battleships on the Hudson at four-minute intervals. Germany believed that it had been done by a submarine, and immediately declared war. It was six months after Gluck's confession that she returned the Philippines and Hawaii to the United States.

In the meanwhile Emil Gluck, the **malevolent** wizard and arch-hater, travelled his **whirlwind** path of destruction. He left no traces. Scientifically thorough, he always cleaned up after himself. His method was to rent a room or a house, and secretly to install his apparatus - which apparatus, by the way, he so perfected and simplified that it occupied little space. After he had accomplished his purpose he carefully removed the apparatus. He bade fair to live out a long life of horrible crime.

The **epidemic** of shooting of New York City policemen was a remarkable affair. It became one of the horror mysteries of the time. In two short weeks over a hundred policemen were shot in the legs by their own revolvers. Inspector Jones did not solve the mystery, but it was his idea that finally outwitted Gluck. On his recommendation the policemen ceased carrying revolvers, and no more accidental shootings occurred.

It was in the early spring of 1940 that Gluck destroyed the Mare Island navy yard. From a room in Vallejo he sent his electric discharges across the Vallejo Straits to Mare Island. He first played his flashes on the battleship Maryland. She lay at the dock of one of the mine magazines. On her forward deck, on a huge temporary platform of timbers, were **disposed** over a hundred mines. These mines were for the defence of the Golden Gate. Any one of these mines was capable of destroying a dozen battleships, and there were over a hundred mines. The destruction was terrific, but it was only Gluck's overture. He played his flashes down the Mare Island shore, blowing up five torpedo boats, the torpedo station, and the great magazine at the eastern end of the island. Returning westward

Crew – *Squad*
Conspiracy – *Plot*
Malevolent – *Malicious*
Whirlwind – *Rapid*
Epidemic – *Widespread occurrence of a disease*
Dispose – *Position*

again, and scooping in occasional **isolated** magazines on the high ground back from the shore, he blew up three cruisers and the battleships Oregon, Delaware, New Hampshire, and Florida - the latter had just gone into dry dock, and the magnificent dry dock was destroyed along with her.

It was a frightful **catastrophe**, and a shiver of horror passed through the land. But it was nothing to what was to follow. In the late fall of that year Emil Gluck made a clean sweep of the Atlantic seaboard from Maine to Florida. Nothing escaped. Forts, mines, coast defences of all sorts, torpedo stations, magazines - everything went up. Three months afterward, in midwinter, he smote the north shore of the Mediterranean from Gibraltar to Greece in the same stupefying manner. A wail went up from the nations. It was clear that human agency was behind all this destruction, and it was equally clear, through Emil Gluck's impartiality, that the destruction was not the work of any particular nation. One thing was patent, namely, that whoever was the human behind it all, that human was a **menace** to the world. No nation was safe. There was no defence against this unknown and all-powerful **foe**. Warfare was futile - nay, not merely futile but itself the very essence of the peril. For a twelve-month the manufacture of powder ceased, and all soldiers and sailors were withdrawn from all fortifications and war vessels. And even a world-disarmament was seriously considered at the Convention of the Powers, held at The Hague at that time.

And then Silas Bannerman, a secret service agent of the United States, leaped into world fame by arresting Emil Gluck. At first Bannerman was laughed at, but he had prepared his case well, and in a few weeks the most sceptical were convinced of Emil Gluck's guilt. The one thing, however, that Silas Bannerman never succeeded in explaining, even to his own satisfaction, was how first he came to connect Gluck with the **atrocious** crimes. It is true, Bannerman was in Vallejo, on secret government business, at the time of the destruction of Mare Island; and it is true that on the streets of Vallejo Emil Gluck was pointed out to him as a queer crank; but no impression was made at the time. It was not until afterward, when on a vacation in the Rocky Mountains and when reading the first published reports of the destruction along the Atlantic Coast, that suddenly Bannerman thought of Emil Gluck. And on the instant there flashed into his mind the connection

Isolate – *Separate*
Catastrophe – *Disaster*
Menace – *Threat*
Foe – *Enemy*
Atrocious – *Terrible*

between Gluck and the **destruction**. It was only an hypothesis, but it was sufficient. The great thing was the **conception** of the hypothesis, in itself an act of unconscious cerebration - a thing as unaccountable as the flashing, for instance, into Newton's mind of the principle of gravitation.

The rest was easy. Where was Gluck at the time of the destruction along the Atlantic sea-board? was the question that formed in Bannerman's mind. By his own request he was put upon the case. In no time he **ascertained** that Gluck had himself been up and down the Atlantic Coast in the late fall of 1940. Also he ascertained that Gluck had been in New York City during the epidemic of the shooting of police officers. Where was Gluck now? was Bannerman's next query. And, as if in answer, came the wholesale destruction along the Mediterranean. Gluck had sailed for Europe a month before - Bannerman knew that. It was not necessary for Bannerman to go to Europe. By means of cable messages and the co-operation of the European secret services, he traced Gluck's course along the Mediterranean and found that in every instance it **coincided** with the blowing up of coast defences and ships. Also, he learned that Gluck had just sailed on the Green Star liner Plutonic for the United States.

The case was complete in Bannerman's mind, though in the interval of waiting he worked up the details. In this he was ably assisted by George Brown, an operator employed by the Wood's System of Wireless Telegraphy. When the Plutonic arrived off Sandy Hook she was boarded by Bannerman from a Government tug, and Emil Gluck was made a prisoner. The trial and the confession followed. In the confession Gluck professed **regret** only for one thing, namely, that he had taken his time. As he said, had he dreamed that he was ever to be discovered he would have worked more rapidly and accomplished a thousand times the destruction he did. His secret died with him, though it is now known that the French Government managed to get access to him and offered him a billion francs for his invention wherewith he was able to direct and closely to confine electric discharges. "What!" was Gluck's reply - "to sell to you that which would enable you to enslave and maltreat suffering Humanity?" And though the war departments of the nations have continued to experiment in their secret laboratories, they have so far failed to light upon the slightest trace of the secret. Emil Gluck was executed

Destruction – *Devastation*
Conception – *Beginning*
Ascertain – *Determine*
Coincide – *Overlap*
Regret – *Remorse*

on December 4, 1941, and so died, at the age of forty-six, one of the world's most **unfortunate** geniuses, a man of **tremendous intellect**, but whose mighty powers, instead of making toward good, were so twisted and warped that he became the most amazing of criminals.

Culled from Mr. A. G. Burnside's "Eccentricitics of Crime," by kind permission of the publishers, Messrs. Holiday and Whitsund.

Food For Thought

Why was Emil Gluck maligned, persecuted and misunderstood by the world? How do you think his love for Irene Tackley ruined him further? Who killed Irene Tackley and John Hartwell? Emil was executed for all his crimes. How did an intellectual like him become an extraordinary criminal?

Tremendous –
Wonderful, Fabulous
Intellect – *Intelligence*
Unfortunate –
Unlucky

An understanding

Q. 1. What is the plot summary of this story? Who was Emil Gluck and why did he become morbid and horrible from his childhood?

Ans. _____

_____

_____

_____

_____

Q. 2. Who brought up Emil and how was he treated in his early, formative years? Give an instance of his childhood days briefly in your own words.

Ans. _____

_____

_____

_____

Q. 3. How was Emil as a student? What happened when he was 27 years old? What is 'Sex and Progress' and how is it treated today?

Ans. _____

_____

_____

_____

Q. 4. What happened after five years of the publication of his famous book, 'Sex and Progress'? How were the newspapers responsible for his miseries?

Ans. _____

_____

_____

_____

Théophile Gautier

Born on August 30, 1811
Died on October 23, 1872
Notable Works: *Une Larme du diable* (1839) ("The Devil's Tear") was written shortly after Gautier's trip to Belgium in 1836. *Le Tricorne enchanté* (1845; "The Magic Hat"), La Fausse Conversion (1846) ("The False Conversion") and many more.

Early Life

Pierre Jules Théophile Gautier was born on August 30, 1811. He was a renowned French poet, dramatist, novelist, journalist, art critic and literary critic.

Gautier was born on August 30, 1811, in Tarbes, capital of Hautes-Pyrénées département in southwestern France. His father, Pierre Gautier, was a fairly cultured minor government official and his mother was Antoinette-Adelaïde Concarde. The family moved to Paris in 1814, taking up residence in the ancient Marais district.

Gautier's education commenced at the prestigious Collège Louis-le-Grand in Paris (fellow alumni include Voltaire and Charles Baudelaire), which he attended for three months before being brought home due to illness. Although he completed the remainder of his education at Collège Charlemagne (alumni include Charles Augustin Sainte-Beuve), Gautier's most significant instruction came from his father, who prompted him to become a Latin scholar by age of 18.

In the aftermath of the 1830 Revolution, Gautier's family experienced hardship and was forced to move to the outskirts of Paris. Deciding to experiment with his own independence and freedom, Gautier chose to stay with friends in the Doyenné district of Paris, living a rather pleasant Bohemian life. Absorbed by the 1848 Revolution, Gautier wrote almost one hundred articles, equivalent to four large books, within nine months in 1848.

Literary Works and Achievements

While Gautier was an ardent defender of Romanticism, his work is difficult to classify and remains a point of reference for many subsequent literary traditions, such as Parnassianism, Symbolism, Decadence and Modernism. He was widely esteemed by writers as diverse as Balzac, Baudelaire, the Goncourt brothers, Flaubert, Proust and Oscar Wilde.

Some of his well-known poems include: *Un Voyage en Espagne* (1843), *La Juive de Constantine* (1846), *Regardez mais ne touchez pas* (1847), *Pierrot en Espagne* (1847), *L'Amour souffle où il veut* (1850) — not completed, *Une Larme du diable* (1839), etc. Some of his famous novels are *Mademoiselle de Maupin* (1835), *Le Roman de*

La Momie (1858) *Le Capitaine Fracasse* (1863), etc. One of his major short stories is *La Morte Amoureuse* (1836) - a classic tale of the supernatural in which a priest receives nocturnal visitations from a female vampire. *Une Larme du diable* (1839) ("The Devil's Tear") was written shortly after Gautier's trip to Belgium in 1836. *Le Tricorne enchanté* (1845; "The Magic Hat"), La Fausse Conversion (1846) ("The False Conversion"), etc are some of his notable works.

Writing Style

In many of Gautier's works, the subject is less important than the pleasure of telling the story. He favored a provocative yet refined style. Gautier did not consider himself to be dramatist but more of a poet and storyteller. His plays were limited because of the time in which he lived. During the Revolution of 1848, many theatres were closed down and therefore plays were scarce.

Later Years

Elected in 1862 as chairman of the Société Nationale des Beaux-Arts, he was surrounded by a committee of important painters: Eugène Delacroix, Pierre Puvis de Chavannes, Édouard Manet, Albert-Ernest Carrier-Belleuse and Gustave Doré.

During the Franco-Prussian War, Gautier made his way back to Paris upon hearing of the Prussian advance on the capital. He remained with his family throughout the invasion and the aftermath of the Commune, eventually dying on October 23, 1872 due to a long-standing cardiac disease at 61 years of age. He is interred at the Cimetière de Montmartre in Paris.

Trivia

From a 21st century standpoint, Gautier's writings about dance are the most significant of his writings. The American writer, Edwin Denby, widely considered the most significant writer about dance in the 20th century, called him "by common consent the greatest of ballet critics".

Through his authorship of the scenario of the ballet, *Giselle*, one of the foundation works of the dance repertoire, his influence remains as great among choreographers and dancers as among critics and balletomanes.

The Mummy's Foot

~Theophile Gautier

I had entered, in an **idle** mood, the shop of one of those curiosity vendors, who are called marchands de bric-a-brac in that Parisian argot which is so perfectly unintelligible elsewhere in France.

You have doubtless glanced occasionally through the windows of some of these shops, which have become so numerous now that it is fashionable to buy antiquated furniture, and that every petty stock broker thinks he must have his chambre au moyen age.

There is one thing there which clings **alike** to the shop of the dealer in old iron, the wareroom of the tapestry-maker, the laboratory of the chemist, and the studio of the painter-- in all those gloomy dens where a furtive daylight filters in through the window shutters, the most manifestly ancient thing is dust--the cobwebs are more authentic than the guimp laces; and the old pear-tree furniture on exhibition is actually younger than the mahogany which arrived but yesterday from America.

The warehouse of my bric-a-brac dealer was a veritable Capharnaum; all ages and all nations seemed to have made their rendezvous there; an Etruscan lamp of red clay stood upon a Boule cabinet, with ebony panels, brightly striped by lines of inlaid brass; a duchess of the court of Louis XV nonchalantly extended her fawn-like feet under a massive table of the time of Louis XIII with heavy spiral supports of oak, and carven designs of chimeras and foliage **intermingled**.

Upon the denticulated shelves of several sideboards glittered immense Japanese dishes with red and blue designs relieved by gilded hatching; side by side with enameled works by Bernard Palissy, representing serpents, frogs, and lizards in relief.

From disemboweled cabinets escaped cascades of silver-lustrous Chinese silks and waves of tinsel, which an oblique sunbeam shot through with luminous beads; while portraits of every era, in frames more or less tarnished, smiled through their yellow varnish.

Idle – *Lazy*
Alike – *Similar*
Intermingle – *Interact*

The striped breastplate of a damascened suit of Milanese armor glittered in one corner; Loves and Nymphs of porcelain; Chinese Grotesques, vases of celadon and crackle-ware; Saxon and old Souvres cups encumbered the shelves and nooks of the apartment.

The dealer followed me closely through the **tortuous** way contrived between the piles of furniture; warding off with his hands the hazardous sweep of my coat-skirts; watching my elbows with the uneasy attention of an antiquarian and a usurer.

It was a singular face that of the merchant--an immense skull, polished like a knee, and surrounded by a thin aureole of white hair, which brought out the clear salmon tint of his complexion all the more strikingly, lent him a false aspect of patriarchal bonhomie, counteracted, however, by the scintillation of two little yellow eyes which trembled in their orbits like two louis-d' or upon quicksilver. The curve of his nose presented an aquiline silhouette, which suggested the Oriental or Jewish type. His hands--thin, slender, full of nerves which projected like strings upon the finger-board of a violin, and armed with claws like those on the terminations of bats' wings--shook with **senile** trembling; but those convulsively agitated hands became firmer than steel pincers or lobsters' claws when they lifted any precious article--an onyx cup, a Venetian glass, or a dish of Bohemian crystal. This strange old man had an aspect so thoroughly rabbinical and cabalistic that he would have been burnt on the mere testimony of his face three centuries ago.

"Will you not buy something from me today, sir? Here is a Malay kreese with a blade **undulating** like flame. Look at those grooves **contrived** for the blood to run along, those teeth set backwards so as to tear out the entrails in withdrawing the weapon--it is a fine character of ferocious arm, and will look well in your collection. This two-handed sword is very beautiful--it is the work of Josepe de la Hera; and this colichemarde, with its fenestrated guard--what a superb **specimen** of handicraft!"

"No; I have quite enough weapons and instruments of carnage;--I want a small figure, something which will suit me as a paper-weight; for I cannot endure those trumpery bronzes which the stationers sell, and which may be found on everybody's desk."

Torturous – *Physical and Mental Pain*
Senile – *Mentally or Physically Weak*
Undulating – *Rolling*
Contrive – *Arrange*
Specimen – *Example*

The old gnome **foraged** among his ancient wares, and finally arranged before me some antique bronzes--so-called, at least; fragments of malachite; little Hindoo or Chinese idols--a kind of poussah toys in jadestone, representing the incarnations of Brahma or Vishnoo, and wonderfully appropriate to the very undivine office of holding papers and letters in place.

I was hesitating between a porcelain dragon, all constellated with warts--its mouth formidable with bristling tusks and ranges of teeth--and an abominable little Mexican fetish, representing the god Zitziliputzili au naturel, when I caught sight of a charming foot, which I at first took for a fragment of some antique Venus.

It had those beautiful ruddy and tawny tints that lend to Florentine bronze that warm living look so much preferable to the gray-green aspect of common bronzes, which might easily be mistaken for statues in a state of putrefaction: satiny gleams played over its rounded forms, doubtless polished by the amorous kisses of twenty centuries; for it seemed a Corinthian bronze, a work of the best era of art--perhaps molded by Lysippus himself.

"That foot will be my choice," I said to the merchant, who regarded me with an **ironical** and **saturnine** air, and held out the object desired that I might examine it more fully.

I was surprised at its lightness; it was not a foot of metal, but in sooth a foot of flesh--an embalmed foot--a mummy's foot. On examining it still more closely the very grain of the skin, and the almost **imperceptible** lines impressed upon it by the texture of the bandages, became perceptible. The toes were slender and delicate, and terminated by perfectly formed nails, pure and transparent as agates; the great toe, slightly separated from the rest, afforded a happy contrast, in the antique style, to the position of the other toes, and lent it an aerial lightness--the grace of a bird's foot;--the sole, scarcely streaked by a few almost imperceptible cross lines, afforded evidence that it had never touched the bare ground, and had only come in contact with the finest matting of Nile rushes, and the softest carpets of panther skin.

"Ha, ha!--you want the foot of the Princess Hermonthis,"-- exclaimed the merchant, with a strange giggle, fixing his owlish eyes upon me--"ha, ha, ha!--for a paper-weight!--an original idea!--artistic idea! Old Pharaoh would certainly

Forage – *Fodder*
Ironical – *Sarcastic*
Saturnine –
Melancholy, Gloomy
Imperceptible –
Unnoticeable

have been surprised had someone told him that the foot of his adored daughter would be used for a paper-weight after he had had a mountain of granite hollowed out as a receptacle for the triple coffin, painted and gilded--covered with hiero-glyphics and beautiful paintings of the Judgment of Souls,"--continued the queer little merchant, half **audibly**, as though talking to himself!

"How much will you charge me for this mummy fragment?"

"Ah, the highest price I can get; for it is a superb piece. If I had the match of it you could not have it for less than five hundred francs;--the daughter of a Pharaoh! Nothing is more rare."

"Assuredly that is not a common article; but, still, how much do you want? In the first place let me warn you that all my wealth consists of just five louis. I can buy anything that costs five louis, but nothing dearer;--you might search my vest pockets and most secret drawers without even finding one poor--five-franc piece more."

"Five louis for the foot of the Princess Hermonthis! Yhat is very little, very little indeed; 'tis an authentic foot," muttered the merchant, shaking his head, and **imparting** a **peculiar** rotary motion to his eyes.

"Well, take it, and I will give you the bandages into the bargain," he added, wrapping the foot in an ancient damask rag--"Very fine! real damask--Indian damask which has never been redyed; it is strong, and yet it is soft," he mumbled, stroking the frayed tissue with his fingers, through the trade-acquired habit which moved him to **praise** even an object of so little value that he himself deemed it only worth the giving away.

He poured the gold coins into a sort of medieval alms-purse hanging at his belt, repeating,

"The foot of the Princess Hermonthis, to be used for a paper-weight!"

Then turning his phosphorescent eyes upon me, he exclaimed in a voice strident as the crying of a cat which has swallowed a fish-bone,

"Old Pharaoh will not be well pleased; he loved his daughter--the dear man!"

"You speak as if you were a contemporary of his. You are old enough, goodness knows! But you do not date back to the

Audible – *Distinct*
Impart – *Inform*
Peculiar – *Odd*
Praise – *Admiration*

Pyramids of Egypt," I answered, laughingly, from the threshold. I went home, delighted with my acquisition.

With the idea of putting it to profitable use as soon as possible, I placed the foot of the divine Princess Hermonthis upon a heap of papers scribbled over with verses, in themselves an undecipherable mosaic work of erasures; articles freshly begun; letters forgotten, and posted in the table drawer instead of the letter-box--an error to which absent-minded people are peculiarly liable. The effect was charming, **bizarre**, and romantic.

Well satisfied with this embellishment, I went out with the gravity and price becoming one who feels that he has the ineffable advantage over all the passers-by whom he elbows, of possessing a piece of the Princess Hermonthis, daughter of Pharaoh.

I looked upon all who did not possess, like myself, a paper-weight so authentically Egyptian, as very ridiculous people; and it seemed to me that the proper occupation of every sensible man should consist in the mere fact of having a mummy's foot upon his desk.

Happily I met some friends, whose presence distracted me in my **infatuation** with this new acquisition. I went to dinner with them; for I could not very well have dined with myself.

When I came back that evening, with my brain slightly confused by a few glasses of wine, a **vague** whiff of Oriental perfume delicately titillated my olfactory nerves. The heat of the room had warmed the natron, bitumen, and myrrh in which the paraschistes, who cut open the bodies of the dead, had bathed the corpse of the princess--it was a perfume at once sweet and penetrating--a perfume that four thousand years had not been able to dissipate.

The Dream of Egypt was Eternity. Her odors have the solidity of granite, and **endure** as long.

I soon drank deeply from the black cup of sleep, for a few hours all remained **opaque** to me; Oblivion and Nothingness inundated me with their somber waves.

Yet light gradually dawned upon the darkness of my mind; dreams commenced to touch me softly in their silent flight.

The eyes of my soul were opened; and I beheld my chamber as it actually was; I might have believed myself awake, but for a vague consciousness which assured me that I slept, and that something fantastic was about to take place.

Bizarre – *Unusual*
Infatuation
– *Momentary Passion*
Vague – *Unclear*
Endure – *Suffer*
Opaque – *Dense*

The odor of the myrrh had **augmented** in intensity; and I felt a slight headache, which I very naturally attributed to several glasses of champagne that we had drunk to the unknown gods and our future fortunes.

I peered through my room with a feeling of expectation which I saw nothing to justify. Every article of furniture was in its proper place; the lamp, softly shaded by its globe of ground crystal, burned upon its bracket; the water-color sketches shone under their Bohemian glass; the curtains hung down languidly; everything wore an aspect of **tranquil slumber**.

After a few moments, however, all this calm interior appeared to become disturbed; the woodwork cracked stealthily; the ash-covered log suddenly emitted a jet of blue flame; and the disks of the pateras seemed like great metallic eyes, watching, like myself, for the things which were about to happen.

My eyes accidentally fell upon the desk where I had placed the foot of the Princess Hermonthis. Instead of remaining quiet--as behooved a foot which had been embalmed for four thousand years--it commenced to act in a nervous manner; contracted itself, and leaped over the papers like a startled frog--one would have imagined that it had suddenly been brought into contact with a galvanic battery. I could distinctly hear the dry sound made by its little heel, hard as the hoof of a gazelle.

I became rather discontented with my acquisition, inasmuch as I wished my paper-weights to be of a sedentary disposition, and thought it very unnatural that feet should walk about without legs; and I commenced to experience a feeling closely akin to fear.

Suddenly I saw the folds of my bed-curtain stir; and heard a bumping sound, like that caused by some person hopping on one foot across the floor. I must **confess** I became alternately hot and cold; that I felt a strange wind chill my back; and that my suddenly rising hair caused my nightcap to execute a leap of several yards.

The bed-curtains opened and I beheld the strangest figure imaginable before me.

It was a young girl of a very deep coffee-brown complexion, like the bayadere Amani, and possessing the purest

Augment – *Enlarge*
Tranquil – *Calm*
Slumber – *Sleep*
Confess – *Admit*

Egyptian type of perfect beauty. Her eyes were almond-shaped and **oblique**, with eyebrows so black that they seemed blue; her nose was **exquisitely** chiseled, almost Greek in its delicacy of outline; and she might indeed have been taken for a Corinthian statue of bronze, but for the prominence of her cheek bones and the slightly African fulness of her lips, which compelled one to recognize her as belonging beyond all doubt to the hieroglyphic race which dwelt upon the banks of the Nile.

Her arms, slender and spindle-shaped, like those of very young girls, were encircled by a peculiar kind of metal bands and bracelets of glass beads; her hair was all twisted into little cords; and she wore upon her bosom a little idol figure of green paste, bearing a whip with seven lashes, which proved it to be an image of Isis. Her brow was adorned with a shining plate of gold; and a few traces of paint relieved the coppery tint of her cheeks.

As for her costume, it was very odd indeed. Fancy a pagne or skirt all formed of little strips of material bedizened with red and black hieroglyphics, stiffened with bitumen, and apparrently belonging to a freshly unbandaged mummy.

In one of those sudden flights of thought so common in dreams I heard the hoarse falsetto of the bric-a-brac dealer, repeating like a monotonous **refrain** the phrase he had uttered in his shop with so enigmatical an intonation,

"Old Pharaoh will not be well pleased. He loved his daughter, the dear man!"

One strange circumstance, which was not at all calculated to **restore** my equanimity, was that the apparition had but one foot; the other was broken off at the ankle!

She approached the table where the foot was starting and fidgeting about more than ever, and there supported herself upon the edge of the desk. I saw her eyes fill with pearly-gleaming tears.

Although she had not as yet spoken, I fully comprehended the thoughts which **agitated** her. She looked at her foot--it was indeed her own--with an exquisitely graceful expression of coquettish sadness; but the foot leaped and ran hither and thither, as though **impelled** on steel springs.

Twice or thrice she extended her hand to seize it, but could not succeed.

Then commenced between the Princess Hermonthis and her foot--which appeared to be **endowed** with a special life of its own--a very fantastic dialogue in a most ancient Coptic tongue,

Oblique – *Slanting*
Exquisite – *Attractive*
Refrain – *Abstain*
Restore – *Reinstate*
Agitate – *Disturb*
Impel – *Compel*

such as might have been spoken thirty centuries ago in the syrinxes of the land of Ser. Luckily, I understood Coptic perfectly well that night.

The Princess Hermonthis cried, in a voice sweet and vibrant as the tones of a crystal bell,

"Well, my dear little foot, you always flee from me; yet I always took good care of you. I bathed you with perfumed water in a bowl of alabaster; I smoothed your heel with pumice-stone mixed with palm oil; your nails were cut with golden scissors and polished with a hippopotamus tooth; I was careful to select tatbebs for you, painted and embroidered and turned up at the toes, which were the **envy** of all the young girls in Egypt. You wore on your great toe rings bearing the device of the sacred Scarabs; and you supported one of the lightest bodies that a lazy foot could **sustain**."

The foot replied, in a pouting and chagrined tone,

"You know well that I do not belong to myself any longer--I have been bought and paid for; the old merchant knew what he was about; he bore you a **grudge** for having refused to espouse him;--this is an ill turn which he has done you. The Arab who violated your royal coffin in the subterranean pit of the necropolis of Thebes was sent thither by him: he desired to prevent you from being present at the **reunion** of the shadowy nations in the cities below. Have you five pieces of gold for my ransom?"

"Alas, no!--my jewels, my rings, my purses of gold and silver, they were all stolen from me," answered the Princess Hermonthis, with a sob.

"Princess," I then exclaimed, "I never retained anybody's foot unjustly;--even though you have not got the five louis which it cost me, I present it to you gladly. I should feel unutterably wretched to think that I were the cause of so **amiable** a person as the Princess Hermonthis being lame."

I delivered this discourse in a royally gallant, troubadour tone, which must have astonished the beautiful Egyptian girl.

She turned a look of deepest gratitude upon me; and her eyes shone with bluish gleams of light.

Endow – *Award*
Envy – *Jealousy*
Sustain – *Endure*
Grudge – *Complaint*
Reunion – *Gathering*
Amiable – *Friendly*

She took her foot--which surrendered itself willingly this time--like a woman about to put on her little shoe, and adjusted it to her leg with much skill.

This operation over, she took a few steps about the room, as though to **assure** herself that she was really no longer lame.

"Ah, how pleased my father will be!--he who was so unhappy because of my mutilation, and who from the moment of my birth set a whole nation at work to hollow me out a tomb so deep that he might preserve me intact until that last day, when souls must be weighed in the balance of Amenthi! Come with me to my father--he will receive you kindly; for you have given me back my foot."

I thought this proposition natural enough. I **arrayed** myself in a dressing gown of large-flowered pattern, which lent me a very Pharaonic aspect; hurriedly put on a pair of Turkish slippers, and informed the Princess Hermonthis that I was ready to follow her.

Before starting, Hermonthis took from her neck the little idol of green paste, and laid it on the scattered sheets of paper which covered the table.

"It is only fair," she observed smilingly, "that I should replace your paper-weight."

She gave me her hand, which felt soft and cold, like the skin of a serpent; and we departed.

We passed for some time with the velocity of an arrow through a fluid and grayish expanse, in which half-formed silhouettes flitted swiftly by us, to right and left.

For an instant we saw only sky and sea.

A few moments later obelisks commenced to tower in the distance. Pylons and vast flights of steps guarded by sphinxes became clearly outlined against the horizon.

We had reached our destination. The princess conducted me to the mountain of rose-colored granite, in the face of which appeared an opening so narrow and low that it would have been difficult to distinguish it from the **fissures** in the rock, had not its location been marked by two steel wrought with sculptures.

Hermonthis kindled a torch, and led the way before me.

We **traversed** corridors hewn through the living rock. Their walls, covered with hieroglyphics and paintings of allegorical processions, might well have occupied thousands of arms for thousands of years in their formation--these corridors, of interminable length, opened into square chambers, in the midst of which pits had been **contrived**, through which we descended by cramp-irons or spiral stairways--these pits again conducted us into other chambers, opening into other corridors, likewise decorated with painted sparrow-hawks, serpents coiled in circles, the symbols of the tau and pedum--

Assure – *Promise*
Array – *Collection*
Fissure – *Crack*
Traverse – *Cross*

prodigious works of art which no living eye can ever examine--interminable legends of granite which only the dead have time to read through all eternity.

At last we found ourselves in a hall so vast, so enormous, so immeasurable, that the eye could not reach its limits; files of monstrous columns stretched far out of sight on every side, between which twinkled livid stars of yellowish flame-- points of light which revealed further depths incalculable in the darkness beyond.

The Princess Hermonthis still held my hand, and graciously saluted the mummies of her acquaintance. My eyes became accustomed to the dim twilight, and objects became **discernible**.

I beheld the kings of the subterranean races seated upon thrones--grand old men, though dry, withered, wrinkled like parchment, and blackened with naphtha and bitumen--all wearing pshents of gold, and breastplaces and gorgets glittering with precious stones; their eyes immovably fixed like the eyes of sphinxes, and their long beards whitened by the snow of centuries. Behind them stood their peoples, in the stiff and constrained posture enjoined by Egyptian art, all eternally preserving the attitude prescribed by the hieratic code. Behind these nations, the cats, ibises, and crocodiles contemporary with them--rendered monstrous of aspect by their swathing bands--mewed, flapped their wings, or extended their jaws in a saurian giggle.

All the Pharaohs were there--Cheops, Chephrenes, Psammetichus, Sesostris, Amenotaph--all the dark rulers of the pyramids and syrinxes--on yet higher thrones sat Chronos and Xixouthros--who was contemporary with the deluge; and Tubal Cain, who reigned before it.

The beard of King Xixouthros had grown seven times around the granite table, upon which he leaned, lost in deep reverie--and buried in dreams.

Further back, through a dusty cloud, I beheld dimly the seventy-two pre-Adamite Kings, with their seventy-two people--forever passed away.

After permitting me to gaze upon this bewildering spectacle a few moments, the Princess Hermonthis presented me to her father Pharaoh, who favored me with a most gracious nod.

"I have found my foot again!--I have found my foot!" cried the Princess, clapping her little hands together with every sign of frantic joy, "It was this gentleman who restored it to me."

contrive –
Manufacture
prodigious –
Abnormal
discernible –
Apparent
Reign – *Rule*

The races of Kemi, the races of Nahasi--all the black, bronzed, and copper-colored nations repeated in chorus,

"The Princess Hermonthis has found her foot again!"

Even Xixouthros himself was visibly affected.

He raised his heavy eyelids, stroked his mustache with his fingers, and turned upon me a glance weighty with centuries.

"By Oms, the dog of Hell, and Tmei, daughter of the Sun and of Truth! This is a brave and worthy lad!" exclaimed Pharaoh, pointing to me with his scepter, which was terminated with a lotus flower.

"What recompense do you desire?"

Filled with that daring inspired by dreams in which nothing seems impossible, I asked him for the hand of the Princess Hermonthis--the hand seemed to me a very proper antithetic recompense for the foot.

Pharaoh opened wide his great eyes of glass in astonishment at my witty request.

"What country do you come from? And what is your age?"

"I am a Frenchman; and I am twenty-seven years old, venerable Pharaoh."

"--Twenty-seven years old! And he wishes to espouse the Princess Hermonthis, who is thirty centuries old!" cried out at once all the Thrones and all the Circles of Nations.

Only Hermonthis herself did not seem to think my request unreasonable.

"If you were even only two thousand years old," replied the ancient King, "I would willingly give you the Princess; but the disproportion is too great; and, besides, we must give our daughters husbands who will last well. You do not know how to preserve yourselves any longer; even those who died only fifteen centuries ago are already no more than a handful of dust--behold! my flesh is solid as basalt; my bones are bars of steel!

"I shall be present on the last day of the world, with the same body and the same features which I had during my lifetime. My daughter Hermonthis will last longer than a statue of bronze.

"Then the last particles of your dust will have been scattered abroad by the winds; and even Isis herself, who was able to find the atoms of Osiris, would scarce be able to recompose your being.

Permit – *Authorize*
Gracious – *Cordial*
Astonishment – *Surprise*
Preserve – *Protect*

"See how vigorous I yet remain, and how mighty is my grasp," he added, shaking my hand in the English fashion with a strength that buried my rings in the flesh of my fingers.

He squeezed me so hard that I awoke, and found my friend Alfred shaking me by the arm to make me get up.

"O you everlasting sleeper!--must I have you carried out into the middle of the street, and fireworks exploded in your ears? It is after noon; don't you recollect your promise to take me with you to see M. Aguado's Spanish pictures?"

"God! I forgot all, all about it," I answered, dressing myself hurriedly; "we will go there at once; I have the permit lying on my desk."

I started to find it;--but fancy my astonishment when I beheld, instead of the mummy's foot I had purchased the evening before, the little green paste idol left in its place by the Princess Hermonthis!

Food For Thought

How and why do you think the mummy's foot disappeared? What was left in its place by the 'Princess Hermonthis? Can you name any other story that you have read similar to this one, or name any other story written by this same author and write its plot summary.

Scarce – *Limited*
Vigorous – *Energetic*
Grasp – *Clutch*
Beheld – *Be believed*

An Understanding

Q. 1. Where does the narrator or the contemporary man venture into? What did he buy from the shop?

Ans. _____

_____

_____

_____

_____

Q. 2. What were the adventures which had befallen on the man who bought a 4000 - year - old foot of Princess Hermonthis?

Ans. _____

_____

_____

_____

_____

Q. 3. Why do you think that the French writer, Theophile Gautier has named the story as 'The Mummy's Foot'? Do you think it is apt or you want to suggest another name for the story? Give reason for your answer.

Ans. _____

_____

_____

_____

_____

Q. 4. Why did the narrator select or prefer to buy the 4000 - year - old foot of Princess Hermonthis? What happened at the end?

Ans. _____

_____

_____

_____

_____

Sir Arthur Conan Doyle

Born on May 22, 1859
Died on July 7, 1930
Notable Works: *Stories of Sherlock Holmes, The Lost World, A Study in Scarlet, etc.*
Honours: Knight Bachelor (1902) and Archie Goodwin Award (2005)

Early Life

Sir Arthur Ignatius Conan Doyle, DL (a Deputy Lieutenant is a military commission in the United Kingdom and one of the several deputies to the Lord Lieutenant of a lieutenancy area) was born on May 22, 1859 at 11 Picardy Place, Edinburgh, Scotland. He was a Scottish physician and writer, most noted for his stories about **the detective, Sherlock Holmes**, generally considered a milestone in the field of crime fiction, and for the **adventures of Professor Challenger**. He was a prolific writer, whose other works include science fiction stories, plays, romances, poetry, non-fiction and historical novels.

His father, Charles Altamont Doyle, was an English of Irish descent, and his mother was an Irish. Although he is now referred to as "Conan Doyle", the origin of this compound surname is uncertain. Supported by wealthy uncles, Conan Doyle was sent to the Roman Catholic Jesuit preparatory school, Hodder Place, Stonyhurst, at the age of nine. He then went on to Stonyhurst College until 1875. From 1875 to 1876, he was educated at the Jesuit school Stella Matutina in Feldkirch, Austria. From 1876 to 1881, he studied medicine at the University of Edinburgh, including a period working in the town of Aston (now a district of Birmingham) and in Sheffield, as well as in Shropshire at Ruyton-XI-Towns. Conan Doyle began writing short stories while studying. His earliest extant fiction, "The Haunted Grange of Goresthorpe", was unsuccessfully submitted to Blackwood's Magazine. His first published piece, "The Mystery of Sasassa Valley", a story set in South Africa, was printed in Chambers's Edinburgh Journal on September 6, 1879. Later that month, on September 20, Sir Arthur Conan Doyle published his first non-fictional article, "Gelsemium as a Poison" in the British Medical Journal.

Following his term at the university, he was employed as a doctor on the Greenland whaler - the Hope of Peterhead in 1880 and after his graduation, as a ship's surgeon on the SS Mayumba during a voyage to the West African coast in 1881. He completed his doctorate on the subject of tabes dorsalis in 1885.

Literary Works and Achievements

His practice was initially not very successful. While waiting for patients, Conan Doyle again began writing stories and composed his first novels, *The Mystery of Cloomber*, not published until 1888, and the unfinished *Narrative of John Smith*, which went unpublished until 2011. He amassed a portfolio of short stories including "The Captain of the Pole-Star" and "J. Habakuk Jephson's Statement", both inspired by Doyle's time at sea.

Doyle struggled to find a publisher for his work. His first significant piece, *A Study in Scarlet*, was taken by Ward Lock & Co on November 20, 1886, giving Doyle £25 for all rights to the story. The piece appeared later that year in the Beeton's Christmas Annual and received good reviews in *The Scotsman and the Glasgow Herald*. The story featured the first appearance of Watson and Sherlock Holmes, partially modelled after his former university teacher, Joseph Bell.

Death of Sherlock Holmes

In December 1893, in order to dedicate more of his time to what he considered his more important works (his historical novels), Conan Doyle had Holmes and Professor Moriarty apparently plunge to their deaths together down the Reichenbach Falls in the story "The Final Problem". Public outcry, however, led him to bring the character back in 1901, in "The Hound of the Baskervilles", though this was set at a time before the Reichenbach incident. In 1903, Conan Doyle published his first Holmes short story in ten years, "The Adventure of the Empty House", in which it was explained that only Moriarty had fallen; but since Holmes had other dangerous enemies—especially, Colonel Sebastian Moran—he had arranged to also be perceived as dead. Holmes ultimately was featured in a total of **56 short stories** and **four Conan Doyle novels**, and has since appeared in many novels and stories by other authors too.

Later Years

Following the death of his wife, Louisa in 1906, the death of his son, Kingsley, just before the end of World War I, and the deaths of his brother, Innes, his two brothers-in-laws (one of whom was E. W. Hornung, creator of the literary character, Raffles) and his two nephews, shortly after the war, Conan Doyle sank into depression. He found solace supporting spiritualism and its attempts to find proof of existence beyond the grave. He was also a member of the renowned paranormal organisation, **The Ghost Club**. Its focus, then and now, is on the scientific study of alleged paranormal activities in order to prove (or refute) the existence of paranormal phenomena.

His book, *The Coming of the Fairies (1921)* shows he was apparently convinced of the veracity of the five Cottingley Fairies photographs (which decades later were exposed as a hoax). *In The History of Spiritualism* (1926), Conan Doyle praised the psychic phenomena and spirit materialisations produced by Eusapia Palladino and Mina "Margery" Crandon.

Sir Arthur Conan Doyle was found clutching his chest in the hall of Windlesham, his house in Crowborough, East Sussex, on July 7, 1930. He died of a heart attack at the age of 71. His grave is at Minstead, England.

Trivia

A statue honours Conan Doyle at Crowborough Cross in Crowborough, where he lived for 23 years. There is also a statue of Sherlock Holmes in Picardy Place, Edinburgh, close to the house, where Conan Doyle was born.

The Terror of Blue John Gap

~ Arthur Conan Doyle

THe following narrative was found among the papers of Dr. James Hardcastle, who died of phthisis on February 4th, 1908, at 36, Upper Coventry Flats, South Kensington. Those who knew him best, while refusing to express an opinion upon this particular statement, are unanimous in asserting that he was a man of a sober and scientific turn of mind, absolutely devoid of imagination, and most unlikely to invent any abnormal series of events. The paper was contained in an envelope, which was docketed, "A Short Account of the Circumstances which occurred near Miss Allerton's Farm in North-West Derbyshire in the Spring of Last Year." The envelope was sealed, and on the other side was written in pencil –

Dear Seaton, -
"It may interest, and perhaps pain you, to know that the incredulity with which you met my story has prevented me from ever opening my mouth upon the subject again. I leave this record after my death, and perhaps strangers may be found to have more confidence in me than my friend."

Inquiry has failed to elicit who this Seaton may have been. I may add that the visit of the deceased to Allerton's Farm, and the general nature of the alarm there, apart from his particular explanation, have been absolutely established. With this foreword I append his account exactly as he left it. It is in the form of a diary, some entries in which have been expanded, while a few have been erased.

April 17. Already I feel the benefit of this wonderful upland air. The farm of the Allertons lies fourteen hundred and twenty feet above sea-level, so it may well be a bracing climate. Beyond the usual morning cough I have very little discomfort, and, what with the fresh milk and the home-grown mutton, I have every chance of putting on weight. I think Saunderson will be pleased.

The two Miss Allertons are charmingly quaint and kind, two dear little hard-working old maids, who are ready to lavish all the heart which might have gone out to husband and to children upon an invalid stranger. Truly, the old maid is a most useful person, one of the reserve forces

Sober – *Abstemious*
Elicit – *Draw out*
Deceased – *Dead*
Quaint – *Old-fashioned*

A Short Account of the Circumstances which occurred near Miss Allerton's Farm in North-West Derbyshire in the Spring of Last Year.

of the community. They talk of the superfluous woman, but what would the poor superfluous man do without her kindly presence? By the way, in their simplicity they very quickly let out the reason why Saunderson recommended their farm. The Professor rose from the ranks himself, and I believe that in his youth he was not above scaring crows in these very fields.

It is a most lonely spot, and the walks are picturesque in the extreme. The farm consists of grazing land lying at the bottom of an irregular valley. On each side are the fantastic limestone hills, formed of rock so soft that you can break it away with your hands. All this country is hollow. Could you strike it with some gigantic hammer it would boom like a drum, or possibly cave in altogether and expose some huge subterranean sea. A great sea there must surely be, for on all sides the streams run into the mountain itself, never to reappear. There are gaps everywhere amid the rocks, and when you pass through them you find yourself in great caverns, which wind down into the bowels of the earth. I have a small bicycle lamp, and it is a perpetual joy to me to carry it into these weird solitudes, and to see the wonderful silver and black effect when I throw its light upon the stalactites which drape the lofty roofs. Shut off the lamp, and you are in the blackest darkness. Turn it on, and it is a scene from the Arabian Nights.

But there is one of these strange openings in the earth which has a special interest, for it is the handiwork, not of nature, but of man. I had never heard of Blue John when I came to these parts. It is the name given to a peculiar mineral of a beautiful purple shade, which is only found at one or two places in the world. It is so rare that an ordinary vase of Blue John would be valued at a great price. The Romans, with that extraordinary instinct of theirs, discovered that it was to be found in this valley, and sank a horizontal shaft deep into the mountain side. The opening of their mine has been called Blue John Gap, a clean-cut arch in the rock, the mouth all overgrown with bushes. It is a goodly passage which the Roman miners have cut, and it intersects some of the great waterworn caves, so that if you enter Blue John Gap you would do well to mark your steps and to have a good store of candles, or you may never make your way back to the daylight again. I have not yet gone deeply into it, but this very day I stood at the mouth of the arched tunnel, and peering down into the

Superfluous – *Surplus*
Perpetual – *Continuous, non-stop*
solitude – *Loneliness*
peculiar – *Odd*
peer – *To peep out or appear slightly*

black recesses beyond, I vowed that when my health returned I would devote some holiday to exploring those mysterious depths and finding out for myself how far the Roman had penetrated into the Derbyshire hills.

Strange how superstitious these countrymen are! I should have thought better of young Armitage, for he is a man of some education and character, and a very fine fellow for his station in life. I was standing at the Blue John Gap when he came across the field to me.

"Well, doctor," said he, "you're not afraid, anyhow."

"Afraid!" I answered. "Afraid of what?"

"Of it," said he, with a jerk of his thumb towards the black vault, "of the terror that lives in the Blue John Cave."

How absurdly easy it is for a legend to arise in a lonely countryside! I examined him as to the reasons for his weird belief. It seems that from time to time sheep have been missing from the fields, carried bodily away, according to Armitage. That they could have wandered away of their own accord and disappeared among the mountains was an explanation to which he would not listen. On one occasion a pool of blood had been found, and some tufts of wool. That also, I pointed out, could be explained in a perfectly natural way. Further, the nights upon which sheep disappeared were invariably very dark, cloudy nights with no moon. This I met with the obvious retort that those were the nights which a commonplace sheep-stealer would naturally choose for his work. On one occasion a gap had been made in a wall, and some of the stones scattered for a considerable distance. Human agency again, in my opinion. Finally, Armitage clinched all his arguments by telling me that he had actually heard the Creature - indeed, that anyone could hear it who remained long enough at the Gap. It was a distant roaring of an immense volume. I could not but smile at this, knowing, as I do, the strange reverberations which come out of an underground water system running amid the chasms of a limestone formation. My incredulity annoyed Armitage so that he turned and left me with some abruptness.

And now comes the queer point about the whole business. I was still standing near the mouth of the cave turning over in my mind the various statements of Armitage, and reflecting how readily they could be explained away, when suddenly, from the depth of the tunnel beside me, there issued a most

Devote – *Dedicate*
Absurd – *Ridiculous*
Legend – *Fable*
Abrupt – *Sudden*

extraordinary sound. How shall I describe it? First of all, it seemed to be a great distance away, far down in the bowels of the earth. Secondly, in spite of this suggestion of distance, it was very loud. Lastly, it was not a boom, nor a crash, such as one would associate with falling water or tumbling rock, but it was a high whine, tremulous and vibrating, almost like the whinnying of a horse. It was certainly a most remarkable experience, and one which for a moment, I must admit, gave a new significance to Armitage's words. I waited by the Blue John Gap for half an hour or more, but there was no return of the sound, so at last I wandered back to the farmhouse, rather mystified by what had occurred. Decidedly I shall explore that cavern when my strength is restored. Of course, Armitage's explanation is too absurd for discussion, and yet that sound was certainly very strange. It still rings in my ears as I write.

April 20. In the last three days I have made several expeditions to the Blue John Gap, and have even penetrated some short distance, but my bicycle lantern is so small and weak that I dare not trust myself very far. I shall do the thing more systematically. I have heard no sound at all, and could almost believe that I had been the victim of some hallucination suggested, perhaps, by Armitage's conversation. Of course, the whole idea is absurd, and yet I must confess that those bushes at the entrance of the cave do present an appearance as if some heavy creature had forced its way through them. I begin to be keenly interested. I have said nothing to the Miss Allertons, for they are quite superstitious enough already, but I have bought some candles, and mean to investigate for myself.

I observed this morning that among the numerous tufts of sheep's wool which lay among the bushes near the cavern there was one which was smeared with blood. Of course, my reason tells me that if sheep wander into such rocky places they are likely to injure themselves, and yet somehow that splash of crimson gave me a sudden shock, and for a moment I found myself shrinking back in horror from the old Roman arch. A fetid breath seemed to ooze from the black depths into which I peered. Could it indeed be possible that some nameless thing, some dreadful presence, was lurking down yonder? I should have been incapable of such feelings in the days of my strength, but one grows more nervous and fanciful when one's health is shaken.

Tremulous
– Unsteady
Mystified *– Puzzled*
Victim *– A person/ thing that suffers harm, death*
Fetid *– Rotten*

For the moment I weakened in my resolution, and was ready to leave the secret of the old mine, if one exists, forever unsolved. But tonight my interest has returned and my nerves grown more steady. Tomorrow I trust that I shall have gone more deeply into this matter.

April 22. Let me try and set down as accurately as I can my extraordinary experience of yesterday. I started in the afternoon, and made my way to the Blue John Gap. I confess that my misgivings returned as I gazed into its depths, and I wished that I had brought a companion to share my exploration. Finally, with a return of resolution, I lit my candle, pushed my way through the briars, and descended into the rocky shaft.

It went down at an acute angle for some fifty feet, the floor being covered with broken stone. Thence there extended a long, straight passage cut in the solid rock. I am no geologist, but the lining of this corridor was certainly of some harder material than limestone, for there were points where I could actually see the tool marks which the old miners had left in their excavation, as fresh as if they had been done yesterday. Down this strange, old-world corridor I stumbled, my feeble flame throwing a dim circle of light around me, which made the shadows beyond the more threatening and obscure. Finally, I came to a spot where the Roman tunnel opened into a water-worn cavern - a huge hall, hung with long white icicles of lime deposit. From this central chamber I could dimly perceive that a number of passages worn by the subterranean streams wound away into the depths of the earth. I was standing there wondering whether I had better return, or whether I dare venture farther into this dangerous labyrinth, when my eyes fell upon something at my feet which strongly arrested my attention.

The greater part of the floor of the cavern was covered with boulders of rock or with hard incrustations of lime, but at this particular point there had been a drip from the distant roof, which had left a patch of soft mud. In the very centre of this there was a huge mark - an ill-defined blotch, deep, broad and irregular, as if a great boulder had fallen upon it. No loose stone lay near, however, nor was there anything to account for the impression. It was far too large to be caused by any possible animal, and besides, there was only the one, and the patch of mud was of such a size that no reasonable

Resolution – *Decree*
Accurate – *Precise*
Stumble – *Stagger*
Obscure –
Incomprehensible,
Vague

stride could have covered it. As I rose from the examination of that singular mark and then looked round into the black shadows which hemmed me in, I must confess that I felt for a moment a most unpleasant sinking of my heart, and that, do what I could, the candle trembled in my outstretched hand.

I soon recovered my nerve, however, when I reflected how absurd it was to associate so huge and shapeless a mark with the track of any known animal. Even an elephant could not have produced it. I determined, therefore, that I would not be scared by vague and senseless fears from carrying out my exploration. Before proceeding, I took good note of a curious rock forma-tion in the wall by which I could recognize the entrance of the Roman tunnel. The precaution was very necessary, for the great cave, so far as I could see it, was intersected by passages. Having made sure of my position, and reassured myself by examining my spare candles and my matches, I advanced slowly over the rocky and uneven surface of the cavern.

And now I come to the point where I met with such sudden and desperate disaster. A stream, some twenty feet broad, ran across my path, and I walked for some little distance along the bank to find a spot where I could cross dry-shod. Finally, I came to a place where a single flat boulder lay near the centre, which I could reach in a stride. As it chanced, however, the rock had been cut away and made top-heavy by the rush of the stream, so that it tilted over as I landed on it and shot me into the ice-cold water. My candle went out, and I found myself floundering about in ut-ter and absolute darkness.

I staggered to my feet again, more amused than alarmed by my adventure. The candle had fallen from my hand, and was lost in the stream, but I had two others in my pocket, so that it was of no importance. I got one of them ready, and drew out my box of matches to light it. Only then did I realize my position. The box had been soaked in my fall into the river. It was impossible to strike the matches.

A cold hand seemed to close round my heart as I realized my position. The darkness was opaque and horrible. It was so utter, one put one's hand up to one's face as if to press off something solid. I stood still, and by an effort I steadied myself. I tried to reconstruct in my mind a map of the floor of the cavern as I had last seen it. Alas! The bearings which had impressed themselves upon my mind were high on the wall, and not to be found by touch. Still, I remembered in a general way how the sides were

Tremble – *Shiver*
Vague – *Indistinct*
Disaster – *Tragedy*
Flounder – *Fatter,*
Waver

situated, and I hoped that by groping my way along them I should at last come to the opening of the Roman tunnel. Moving very slowly, and continually striking against the rocks, I set out on this desperate quest.

But I very soon realized how impossible it was. In that black, velvety darkness one lost all one's bearings in an instant. Before I had made a dozen paces, I was utterly bewildered as to my whereabouts. The rippling of the stream, which was the one sound audible, showed me where it lay, but the moment that I left its bank I was utterly lost. The idea of finding my way back in absolute darkness through that limestone labyrinth was clearly an impossible one.

I sat down upon a boulder and reflected upon my unfortunate plight. I had not told anyone that I proposed to come to the Blue John mine, and it was unlikely that a search party would come after me. Therefore I must trust to my own resources to get clear of the danger. There was only one hope, and that was that the matches might dry. When I fell into the river, only half of me had got thoroughly wet. My left shoulder had remained above the water. I took the box of matches, therefore, and put it into my left armpit. The moist air of the cavern might possibly be counteracted by the heat of my body, but even so, I knew that I could not hope to get a light for many hours. Meanwhile there was nothing for it but to wait.

By good luck I had slipped several biscuits into my pocket before I left the farm house. These I now devoured, and washed them down with a draught from that wretched stream which had been the cause of all my misfortunes. Then I felt about for a comfortable seat among the rocks, and, having discovered a place where I could get a support for my back, I stretched out my legs and settled myself down to wait. I was wretchedly damp and cold, but I tried to cheer myself with the reflection that modern science prescribed open windows and walks in all weather for my disease. Gradually, lulled by the monotonous gurgle of the stream, and by the absolute darkness, I sank into an uneasy slumber.

How long this lasted I cannot say. It may have been for an hour, it may have been for several. Suddenly I sat up on my rock couch, with every nerve thrilling and every sense acutely on the alert. Beyond all doubt I had heard a sound - some sound very distinct from the gurgling of the waters. It had passed, but the reverberation of it still lingered in my ear.

Quest – *Mission*
Bewilder – *Confuse*
Audible – *Perceptible*
Plight – *Dilemma*

Was it a search party? They would most certainly have shouted, and vague as this sound was which had wakened me, it was very distinct from the human voice. I sat palpitating and hardly daring to breathe. There it was again! And again! Now it had become continuous. It was a tread - yes, surely it was the tread of some living creature. But what a tread it was! It gave one the impression of enormous weight carried upon sponge-like feet, which gave forth a muffled but ear-filling sound. The darkness was as complete as ever, but the tread was regular and decisive. And it was coming beyond all question in my direction.

My skin grew cold, and my hair stood on end as I listened to that steady and ponderous footfall. There was some creature there, and surely by the speed of its advance, it was one which could see in the dark. I crouched low on my rock and tried to blend myself into it. The steps grew nearer still, then stopped, and presently I was aware of a loud lapping and gurgling. The creature was drinking at the stream. Then again there was silence, broken by a succession of long sniffs and snorts of tremendous volume and energy. Had it caught the scent of me? My own nostrils were filled by a low fetid odour, mephitic and abominable. Then I heard the steps again. They were on my side of the stream now. The stones rattled within a few yards of where I lay. Hardly daring to breathe, I crouched upon my rock. Then the steps drew away. I heard the splash as it returned across the river, and the sound died away into the distance in the direction from which it had come.

For a long time I lay upon the rock, too much horrified to move. I thought of the sound which I had heard coming from the depths of the cave, of Armitage's fears, of the strange impression in the mud, and now came this final and absolute proof that there was indeed some inconceivable monster, something utterly unearthly and dreadful, which lurked in the hollow of the mountain. Of its nature or form I could frame no conception, save that it was both light-footed and gigantic. The combat between my reason, which told me that such things could not be, and my senses, which told me that they were, raged within me as I lay. Finally, I was almost ready to persuade myself that this experience had been part of some evil dream, and that my abnormal condition might have conjured up an hallucination. But there remained one final experience which removed the last possibility of doubt from my mind.

Reverberation
– *Echo*
Tread – *Walk*
Crouch – *Bend*
Conjure – *Summon*

I had taken my matches from my armpit and felt them. They seemed perfectly hard and dry. Stooping down into a crevice of the rocks, I tried one of them. To my delight it took fire at once. I lit the candle, and, with a terrified backward glance into the obscure depths of the cavern, I hurried in the direction of the Roman passage. As I did so I passed the patch of mud on which I had seen the huge imprint. Now I stood astonished before it, for there were three similar imprints upon its surface, enormous in size, irregular in outline, of a depth which indicated the ponderous weight which had left them. Then a great terror surged over me. Stooping and shading my candle with my hand, I ran in a frenzy of fear to the rocky archway, hastened up it, and never stopped until, with weary feet and panting lungs, I rushed up the final slope of stones, broke through the tangle of briars, and flung myself exhausted upon the soft grass under the peaceful light of the stars. It was three in the morning when I reached the farm house, and today I am all unstrung and quivering after my terrific adventure. As yet I have told no one. I must move warily in the matter. What would the poor lonely women, or the uneducated yokels here think of it if I were to tell them my experience? Let me go to someone who can understand and advise.

April 25. I was laid up in bed for two days after my incredible adventure in the cavern. I use the adjective with a very definite meaning, for I have had an experience since which has shocked me almost as much as the other. I have said that I was looking round for someone who could advise me. There is a Dr. Mark Johnson who practices some few miles away, to whom I had a note of recommendation from Professor Saunderson. To him I drove, when I was strong enough to get about, and I recounted to him my whole strange experience. He listened intently, and then carefully examined me, paying special attention to my reflexes and to the pupils of my eyes. When he had finished, he refused to discuss my adventure, saying that it was entirely beyond him, but he gave me the card of a Mr. Picton at Castleton, with the advice that I should instantly go to him and tell him the story exactly as I had done to himself. He was, according to my adviser, the very man who was pre-eminently suited to help me. I went on to the station, therefore, and made my way to the little town, which is some ten miles away. Mr. Picton appeared to be a man of importance, as his brass plate was displayed upon the door of a considerable building on the outskirts of the town. I was about to ring his bell, when some misgiving came into my mind,

Crevice – *A crack, gap*
Obscure – *Incomprehensible, Not clear*
Surge – *Rush forward*
Quivering – *Shaking*

and, crossing to a neighbouring shop, I asked the man behind the counter if he could tell me anything of Mr. Picton. "Why," said he, "he is the best mad doctor in Derbyshire, and yonder is his asylum." You can imagine that it was not long before I had shaken the dust of Castleton from my feet and returned to the farm, cursing all unimaginative pedants who cannot conceive that there may be things in creation which have never yet chanced to come across their mole's vision. After all, now that I am cooler, I can afford to admit that I have been no more sympathetic to Armitage than Dr. Johnson has been to me.

April 27. When I was a student I had the reputation of being a man of courage and enterprise. I remember that when there was a ghost-hunt at Coltbridge it was I who sat up in the haunted house. Is it advancing years (after all, I am only thirty-five), or is it this physical malady which has caused degeneration? Certainly my heart quails when I think of that horrible cavern in the hill, and the certainty that it has some monstrous occupant. What shall I do? There is not an hour in the day that I do not debate the question. If I say nothing, then the mystery remains unsolved. If I do say anything, then I have the alternative of mad alarm over the whole countryside, or of absolute incredulity which may end in consigning me to an asylum. On the whole, I think that my best course is to wait, and to prepare for some expedition which shall be more deliberate and better thought out than the last. As a first step I have been to Castleton and obtained a few essentials - a large acetylene lantern for one thing, and a good double-barrelled sporting rifle for another. The latter I have hired, but I have bought a dozen heavy game cartridges, which would bring down a rhinoceros. Now I am ready for my troglodyte friend. Give me better health and a little spate of energy, and I shall try conclusions with him yet. But who and what is he? Ah! there is the question which stands between me and my sleep. How many theories do I form, only to discard each in turn! It is all so utterly unthinkable. And yet the cry, the footmark, the tread in the cavern - no reasoning can get past these I think of the old-world legends of dragons and of other monsters. Were they, perhaps, not such fairy tales as we have thought? Can it be that there is some fact which underlies them, and am I, of all mortals, the one who is chosen to expose it?

May 3. For several days I have been laid up by the vagaries of an English spring, and during those days there have been developments, the true and sinister meaning of which no one can

Misgiving – *Scruple*
Courage – *Bravery*
Quail – *Recoil*
Legend – *Fable*

Greatest Science Fiction Stories

appreciate save myself. I may say that we have had cloudy and moonless nights of late, which according to my information were the seasons upon which sheep disappeared. Well, sheep have disappeared. Two of Miss Allerton's, one of old Pearson's of the Cat Walk, and one of Mrs. Moulton's. Four in all during three nights. No trace is left of them at all, and the countryside is buzzing with rumours of gipsies and of sheep-stealers.

But there is something more serious than that. Young Armitage has disappeared also. He left his moorland cottage early on Wednesday night and has never been heard of since. He was an unattached man, so there is less sensation than would otherwise be the case. The popular explanation is that he owes money, and has found a situation in some other part of the country, whence he will presently write for his belongings. But I have grave misgivings. Is it not much more likely that the recent tragedy of the sheep has caused him to take some steps which may have ended in his own destruction? He may, for example, have lain in wait for the creature and been carried off by it into the recesses of the mountains. What an inconceivable fate for a civilized Englishman of the twentieth century! And yet I feel that it is possible and even probable. But in that case, how far am I answerable both for his death and for any other mishap which may occur? Surely with the knowledge I already possess it must be my duty to see that something is done, or if necessary to do it myself. It must be the latter, for this morning I went down to the local police station and told my story. The inspector entered it all in a large book and bowed me out with commendable gravity, but I heard a burst of laughter before I had got down his garden path. No doubt he was recounting my adventure to his family.

June 10. I am writing this, propped up in bed, six weeks after my last entry in this journal. I have gone through a terrible shock both to mind and body, arising from such an experience as has seldom befallen a human being before. But I have attained my end. The danger from the terror which dwells in the Blue John Gap has passed never to return. Thus much at least I, a broken invalid, have done for the common good. Let me now recount what occurred as clearly as I may.

Underlie – *Lie beneath*
Sinister – *Bad, evil, Wicked*
Fate – *Destiny*

The night of Friday, May 3rd, was dark and cloudy - the very night for the monster to walk. About eleven o'clock I went from the farm house with my lantern and my rifle, having first left a note upon the table of my bedroom in which I said that, if I were missing, search should be made for me in the direction of the Gap. I made my way to the mouth of the Roman shaft, and, having perched myself among the rocks close to the opening, I shut off my lantern and waited patiently with my loaded rifle ready to my hand.

It was a melancholy vigil. All down the winding valley I could see the scattered lights of the farm houses, and the church clock of Chapel-le-Dale tolling the hours came faintly to my ears. These tokens of my fellow men served only to make my own position seem the more lonely, and to call for a greater effort to overcome the terror which tempted me continually to get back to the farm, and abandon for ever this dangerous quest. And yet there lies deep in every man a rooted self-respect which makes it hard for him to turn back from that which he has once undertaken. This feeling of personal pride was my salvation now, and it was that alone which held me fast when every instinct of my nature was dragging me away. I am glad now that I had the strength. In spite of all that is has cost me, my manhood is at least above reproach.

Twelve o'clock struck in the distant church, then one, then two. It was the darkest hour of the night. The clouds were drifting low, and there was not a star in the sky. An owl was hooting somewhere among the rocks, but no other sound, save the gentle sough of the wind, came to my ears. And then suddenly I heard it! From far away down the tunnel came those muffled steps, so soft and yet so ponderous. I heard also the rattle of stones as they gave way under that giant tread. They drew nearer. They were close upon me. I heard the crashing of the bushes round the entrance, and then dimly through the darkness I was conscious of the loom of some enormous shape, some monstrous inchoate creature, passing swiftly and very silently out from the tunnel. I was paralysed with fear and amazement. Long as I had waited, now that it had actually come I was unprepared for the shock. I lay motionless and breathless, whilst the great dark mass whisked by me and was swallowed up in the night.

But now I nerved myself for its return. No sound came from the sleeping countryside to tell of the horror which was loose. In no way could I judge how far off it was, what it was doing, or when it might be back. But not a second time should my nerve fail me, not a second time should it pass unchallenged. I swore it between my clenched teeth as I laid my cocked rifle across the rock.

Dwell – *Reside*
Perch – *Rest on*
Abandon – *Dump*
Inchoate – *Immature, Undeveloped*

And yet it nearly happened. There was no warning of approach now as the creature passed over the grass. Suddenly, like a dark, drifting shadow, the huge bulk loomed up once more before me, making for the entrance of the cave. Again came that paralysis of volition which held my crooked forefinger impotent upon the trigger. But with a desperate effort I shook it off. Even as the brushwood rustled, and the monstrous beast blended with the shadow of the Gap, I fired at the retreating form. In the blaze of the gun I caught a glimpse of a great shaggy mass, something with rough and bristling hair of a withered grey colour, fading away to white in its lower parts, the huge body supported upon short, thick, curving legs. I had just that glance, and then I heard the rattle of the stones as the creature tore down into its burrow. In an instant, with a triumphant revulsion of feeling, I had cast my fears to the wind, and uncovering my powerful lantern, with my rifle in my hand, I sprang down from my rock and rushed after the monster down the old Roman shaft.

My splendid lamp cast a brilliant flood of vivid light in front of me, very different from the yellow glimmer which had aided me down the same passage only twelve days before. As I ran, I saw the great beast lurching along before me, its huge bulk filling up the whole space from wall to wall. Its hair looked like coarse faded oakum, and hung down in long, dense masses which swayed as it moved. It was like an enormous unclipped sheep in its fleece, but in size it was far larger than the largest elephant, and its breadth seemed to be nearly as great as its height. It fills me with amazement now to think that I should have dared to follow such a horror into the bowels of the earth, but when one's blood is up, and when one's quarry seems to be flying, the old primeval hunting-spirit awakes and prudence is cast to the wind. Rifle in hand, I ran at the top of my speed upon the trail of the monster.

I had seen that the creature was swift. Now I was to find out to my cost that it was also very cunning. I had imagined that it was in panic flight, and that I had only to pursue it. The idea that it might turn upon me never entered my excited brain. I have already explained that the passage down which I was racing opened into a great central cave. Into this I rushed, fearful lest I should lose all trace of the beast. But he had turned upon his own traces, and in a moment we were face to face.

That picture, seen in the brilliant white light of the lantern, is etched for ever upon my brain. He had reared up on his hind legs as a bear would do, and stood above me, enormous, menacing

Volition – *Willing, Choosing*
Retreat – *Move away*
Triumphant – *Successful, Victorious*
Prudence – *Carefulness, Cautiousness*

- such a creature as no nightmare had ever brought to my imagination. I have said that he reared like a bear, and there was something bear-like - if one could conceive a bear which was ten-fold
the bulk of any bear seen upon earth - in his whole pose and attitude, in his great crooked forelegs with their ivory-white claws,
in his rugged skin, and in his red, gaping mouth, fringed with
monstrous fangs. Only in one point did he differ from the bear,
or from any other creature which walks the earth, and even at
that supreme moment a shudder of horror passed over me as I
observed that the eyes which glistened in the glow of my lantern
were huge, projecting bulbs, white and sightless. For a moment
his great paws swung over my head. The next he fell forward
upon me, I and my broken lantern crashed to the earth, and I remember no more.

When I came to myself I was back in the farm house of the
Allertons. Two days had passed since my terrible adventure in
the Blue John Gap. It seems that I had lain all night in the cave
insensible from concussion of the brain, with my left arm and two
ribs badly fractured. In the morning my note had been found,
a search party of a dozen farmers assembled, and I had been
tracked down and carried back to my bedroom, where I had lain
in high delirium ever since. There was, it seems, no sign of the
creature, and no bloodstain which would show that my bullet
had found him as he passed. Save for my own plight and the
marks upon the mud, there was nothing to prove that what I said
was true.

Six weeks have now elapsed, and I am able to sit out once
more in the sunshine. Just opposite me is the steep hillside, grey
with shaly rock, and yonder on its flank is the dark cleft which
marks the opening of the Blue John Gap. But it is no longer a
source of terror. Never again through that ill-omened tunnel shall
any strange shape flit out into the world of men. The educated
and the scientific, the Dr. Johnsons and the like, may smile at
my narrative, but the poorer folk of the countryside had never a
doubt as to its truth. On the day after my recovering consciousness they assembled in their hundreds round the Blue John Gap.
As the Castleton Courier said,

"It was useless for our correspondent, or for any of the ad-
venturous gentlemen who had come from Matlock, Buxton,
and other parts, to offer to descend, to explore the cave to the
end, and to finally test the extraordinary narrative of Dr. James
Hardcastle. The country people had taken the matter into their

Cunning – *Shrewd,
Crafty*
Panic – *Fright*
Insensible – *Numb*
Delirium –
Hallucination
Elapse – *Pass by or
to slip*

own hands, and from an early hour of the morning they had worked hard in stopping up the entrance of the tunnel. There is a sharp slope where the shaft begins, and great boulders, rolled along by many willing hands, were thrust down it until the Gap was absolutely sealed. So ends the episode which has caused such excitement throughout the country. Local opinion is fiercely divided upon the subject. On the one hand are those who point to Dr. Hardcastle's impaired health, and to the possibility of cerebral lesions of tubercular origin giving rise to strange hallucinations. Some idee fixe, according to these gentlemen, caused the doctor to wander down the tunnel, and a fall among the rocks was sufficient to account for his injuries. On the other hand, a legend of a strange creature in the Gap has existed for some months back, and the farmers look upon Dr. Hardcastle's narrative and his personal injuries as a final corroboration. So the matter stands, and so the matter will continue to stand, for no definite solution seems to us to be now possible. It transcends human wit to give any scientific explanation which could cover the alleged facts."

Perhaps before the Courier published these words they would have been wise to send their representative to me. I have thought the matter out, as no one else has occasion to do, and it is possible that I might have removed some of the more obvious difficulties of the narrative and brought it one degree nearer to scientific acceptance. Let me then write down the only explanation which seems to me to elucidate what I know to my cost to have been a series of facts. My theory may seem to be wildly improbable, but at least no one can venture to say that it is impossible.

My view is - and it was formed, as is shown by my diary, before my personal adventure - that in this part of England there is a vast subterranean lake or sea, which is fed by the great number of streams which pass down through the limestone. Where there is a large collection of water there must also be some evaporation, mists or rain, and a possibility of vegetation. This in turn suggests that there may be animal life, arising, as the vegetable life would also do, from those seeds and types which had been introduced at an early period of the world's history, when communication with the outer air was more easy. This place had then developed a fauna and flora of its own, including such monsters as the one which I had seen, which may well have been the old cave bear, enormously enlarged and modified by its new environment. For countless aeons the internal and the external creation had kept

Thrust – *Push*
Impair – *Damage*
Wander – *Stroll*
Transcend – *Rise above*

apart, growing steadily away from each other. Then there had come some rift in the depths of the mountain which had enabled one creature to wander up and, by means of the Roman tunnel, to reach the open air. Like all subterranean life, it had lost the power of sight, but this had no doubt been compensated for by nature in other directions. Certainly it had some means of finding its way about, and of hunting down the sheep upon the hillside. As to its choice of dark nights, it is part of my theory that light was painful to those great white eyeballs, and that it was only a pitch-black world which it could tolerate. Perhaps, indeed, it was the glare of my lantern which saved my life at that awful moment when we were face to face. So I read the riddle. I leave these facts behind me, and if you can explain them, do so; or if you choose to doubt them, do so. Neither your belief nor your incredulity can alter them, nor affect one whose task is nearly over.

So ended the strange narrative of Dr. James Hardcastle.

Food For Thought

What do most of the Londoners believe about Dr. Hardcastle's adventures? Is their belief true or the injuries that Dr. Hardcastle suggest something really alarming and mysterious happening in the Blue John Gap?

Tolerate – *To endure*
Incredulity
– *Disbelief*
Alter – *Change*

An Understanding

Q. 1. What is the plot summary of this story? Why do you think that 'The Terror of Blue John Gap' is an apt heading for the story? If you donot agree, then can you suggest any other title for the story? Support your answer with appropriate reasons.

Ans. _____

_____

_____

_____

_____

Q. 2. Where does the story begin and who is Dr. James Hardcastle? What did he discover and where was it?

Ans. _____

_____

_____

_____

_____

Q. 3. What is 'Blue John' and what was actually done there? Who was Armitage and what did he warn Dr. Hardcastle?

Ans. _____

_____

_____

_____

_____

Q. 4. What happened to Dr. Hardcastle during his investigations in the Blue John Gap? What did the blood trails leading into the Gap indicate?

Ans. _____

_____

_____

_____

_____

Charles Dickens

Born on February 7, 1812(1812-02-07)
Died on June 9, 1870(1870-06-09) (aged 58)
Notable Works: *The Pickwick Papers, Oliver Twist, A Christmas Carol, David Copperfield, Bleak House, Hard Times, A Tale of Two Cities, Great Expectations,* etc.

Early Life

Charles John Huffam Dickens was born on February 7, 1812 – 9 June 1870) was an English writer and social critic who is generally regarded as the greatest novelist of the Victorian period and the creator of some of the world's most memorable fictional characters. During his lifetime Dickens' works enjoyed unprecedented popularity and fame, but it was in the twentieth century that his literary genius was fully recognized by critics and scholars. His novels and short stories continue to enjoy an enduring popularity among the general reading public.

Born in Portsmouth, England, Dickens left school to work in a factory after his father was thrown into debtors' prison. Though he had little formal education, his early impoverishment drove him to succeed. He edited a weekly journal for 20 years, wrote 15 novels and hundreds of short stories and non-fiction articles, lectured and performed extensively, was an indefatigable letter writer, and campaigned vigorously for children's rights, education, and other social reforms.

Literary Works and Achievements

Dickens rocketed to fame with the 1836 serial publication of *The Pickwick Papers*. Within a few years, he had become an international literary celebrity, celebrated for his *humour*, *satire*, and *keen observation* of character and society. His novels, most published in monthly or weekly instalments, pioneered the serial publication of narrative fiction, which became the dominant Victorian mode for novel publication. Dickens was regarded as the 'literary colossus' of his age. His 1843 novella, *A Christmas Carol*, is one of the most influential works ever written, and it remains popular and continues to inspire adaptations in every artistic genre. His creative genius has been praised by fellow writers—from Leo Tolstoy to G. K. Chesterton and George Orwell—for its realism, comedy, prose style, unique characterisations, and social criticism.

Charles Dickens published over a dozen major novels, a large number of short stories (including a number of Christmas-themed stories), a handful of plays, and several non-fiction books. Dickens's novels were initially serialised in weekly and monthly magazines, then reprinted in standard book formats. Some of his notable works are: *The Posthumous Papers of the Pickwick Club, The Adventures of Oliver Twist, The Life and Adventures of Nicholas Nickleby, The Old Curiosity Shop*, etc. Some of his Christmas books: *A Christmas Carol* (1843), *The Chimes* (1844), *The Cricket on the*

Hearth (1845), *The Battle of Life* (1846), *The Haunted Man and the Ghost's Bargain* and many more.

Some of his popular short story collections are: *Sketches by Boz* (1836), *The Mudfog Papers* (1837), *Reprinted Pieces* (1861), *The Uncommercial Traveller (1860–1869),* etc. Some of his selected non-fiction, poetry and plays include: *The Village Coquettes* (play, 1836), *The Fine Old English Gentleman* (poetry, 1841), *Memoirs of Joseph Grimaldi* (1838), *Pictures from Italy* (1846), *The Life of Our Lord: As written for his children* (1849), *A Child's History of England* (1853), *The Frozen Deep* (play, 1857), etc.

Writing Style

Dickens loved the style of the 18th century picaresque novels which he found in abundance on his father's shelves. According to Ackroyd, other than these, perhaps the most important literary influence on him was derived from the fables of *The Arabian Nights*.

His writing style is marked by a profuse linguistic creativity. Satire, flourishing in his gift for caricature is his forte. An early reviewer compared him to Hogarth for his keen practical sense of the ludicrous side of life, though his acclaimed mastery of varieties of class idiom may in fact mirror the conventions of contemporary popular theatre. Dickens worked intensively on developing arresting names for his characters that would reverberate with associations for his readers, and assist the development of motifs in the storyline.

Later Years

On June 9, 1865, while returning from Paris with Ternan, Dickens was involved in the Staplehurst rail crash. The first seven carriages of the train plunged off a cast iron bridge under repair. The only first-class carriage to remain on the track was the one in which Dickens was travelling. Dickens tended and comforted the wounded and the dying, and saved some lives, before rescuers arrived, with a flask of brandy and a hat refreshed with water. Before leaving, he remembered the unfinished manuscript for *Our Mutual Friend*, and he returned to his carriage to retrieve it. Typically, Dickens later used this experience as material for his short ghost story, *The Signal-Man* in which the central character has a premonition of his own death in a rail crash. He based the story around several previous rail accidents, such as the Clayton Tunnel rail crash of 1861. On June 8, 1870, Dickens suffered a second stroke at his home, after a full day's work on Edwin Drood. He never regained consciousness, and the next day, he expired.

Trivia

Charles Dickens was commemorated on the Series E £10 note issued by the Bank of England that was in circulation in the UK between 1992 and 2003. His portrait appeared on the reverse of the note accompanied by a scene from *The Pickwick Papers*. A theme park, *Dickens World*, was built on the site of the former naval dockyard where Dickens's father once worked in the Navy Pay Office, opened in Chatham in 2007.

The Haunted House

~ Charles Dickens

CHAPTER I - THE MORTALS IN THE HOUSE

UNder none of the accredited ghostly circumstances, and environed by none of the conventional ghostly surroundings, did I first make acquaintance with the house which is the subject of this Christmas piece. I saw it in the daylight, with the sun upon it. There was no wind, no rain, no lightning, no thunder, no awful or unwanted circumstance, of any kind, to heighten its effect. More than that, I had come to it direct from a railway station, it was not more than a mile distant from the railway station; and, as I stood outside the house, looking back upon the way I had come, I could see the goods train running smoothly along the embankment in the valley. I will not say that everything was utterly commonplace, because I doubt if anything can be that, except to utterly commonplace people--and there my vanity steps in; but, I will take it on myself to say that anybody might see the house as I saw it, any fine autumn morning.

The manner of my lighting on it was this.

I was travelling towards London out of the North, intending to stop by the way, to look at the house. My health required a temporary residence in the country; and a friend of mine who knew that, and who had happened to drive past the house, had written to me to suggest it as a likely place. I had got into the train at midnight, and had fallen asleep, and had woke up and had sat looking out of window at the brilliant Northern Lights in the sky, and had fallen asleep again, and had woke up again to find the night gone, with the usual discontented conviction on me that I hadn't been to sleep at all; -- upon which question, in the first imbecility of that condition, I am ashamed to believe that I would have done wager by battle with the man who sat opposite me. That opposite man had had, through the night -- as that opposite man always has -- several legs too many, and all of them too long. In addition to this unreasonable conduct (which was only to be expected of him), he had had a pencil and a pocket book, and had been perpetually listening and taking notes. It had appeared to me that these aggravating notes related to the jolts and bumps of

Accredit – *Recognize*
Acquaint – *Explain*
Vanity – *Pride*
Perpetual – *Continuous*

the carriage, and I should have resigned myself to his taking them, under a general supposition that he was in the civil-engineering way of life, if he had not sat staring straight over my head whenever he listened. He was a goggle-eyed gentleman of a perplexed aspect, and his demeanour became unbearable.

It was a cold, dead morning (the sun not being up yet), and when I had out-watched the paling light of the fires of the iron country, and the curtain of heavy smoke that hung at once between me and the stars and between me and the day, I turned to my fellow traveller and said,

"I BEG your pardon, sir, but do you observe anything particular in me?" For, really, he appeared to be taking down, either my travelling cap or my hair, with a minuteness that was a liberty.

The goggle-eyed gentleman withdrew his eyes from behind me, as if the back of the carriage were a hundred miles off, and said, with a lofty look of compassion for my insignificance,

"In you, sir? -- B."

"B, sir?" said I, growing warm.

"I have nothing to do with you, sir," returned the gentleman; "pray let me listen -- O."

He enunciated this vowel after a pause, and noted it down.

At first I was alarmed, for an express lunatic and no communication with the guard, is a serious position. The thought came to my relief that the gentleman might be what is popularly called a Rapper, one of a sect for (some of) whom I have the highest respect, but whom I don't believe in. I was going to ask him the question, when he took the bread out of my mouth.

"You will excuse me," said the gentleman contemptuously, "if I am too much in advance of common humanity to trouble myself at all about it. I have passed the night -- as indeed I pass the whole of my time now -- in spiritual intercourse."

"O!" said I, somewhat snappishly.

"The conferences of the night began," continued the gentleman, turning several leaves of his notebook, "with this message, 'Evil communications corrupt good manners.'"

"Sound," said I; "but, absolutely new?"

"New from spirits," returned the gentleman.

Perplex – *Stun*
Demeanour
– *Manner*
Lofty – *High*
Compassion –
Sympathy

I could only repeat my rather snappish "O!" and ask if I might be favoured with the last communication. "'A bird in the hand,'" said the gentleman, reading his last entry with great solemnity, "'is worth two in the bosh.'"

"Truly I am of the same opinion," said I; "but shouldn't it be bush?"

"It came to me, bosh," returned the gentleman.

The gentleman then informed me that the spirit of Socrates had delivered this special revelation in the course of the night. "My friend, I hope you are pretty well. There are two in this railway carriage. How do you do? There are seventeen thousand four hundred and seventy-nine spirits here, but you cannot see them. Pythagoras is here. He is not at liberty to mention it, but hopes you like travelling." Galileo likewise had dropped in, with this scientific intelligence. "I am glad to see you, AMICO. COME STA? Water will freeze when it is cold enough. ADDIO!" In the course of the night, also, the following phenomena had occurred. Bishop Butler had insisted on spelling his name, "Bubler," for which offence against orthography and good manners he had been dismissed as out of temper. John Milton (suspected of wilful mystification) had repudiated the authorship of Paradise Lost, and had introduced, as joint authors of that poem, two Unknown gentlemen, respectively named Grungers and Scadgingtone. And Prince Arthur, nephew of King John of England, had described himself as tolerably comfortable in the seventh circle, where he was learning to paint on velvet, under the direction of Mrs. Trimmer and Mary Queen of Scots.

If this should meet the eye of the gentleman who favoured me with these disclosures, I trust he will excuse my confessing that the sight of the rising sun, and the contemplation of the magnificent Order of the vast Universe, made me impatient of them. In a word, I was so impatient of them, that I was mightily glad to get out at the next station, and to exchange these clouds and vapours for the free air of Heaven.

By that time it was a beautiful morning. As I walked away among such leaves as had already fallen from the golden, brown, and russet trees; and as I looked around me on the wonders of Creation, and thought of the steady, unchanging, and harmonious laws by which they are sustained; the gentleman's spiritual intercourse seemed to me as poor a piece of

Revelation
– *Disclosure*
Phenomena –
Occurrence
Offence – *Crime*
Repudiate – *Reject*

journey work as ever this world saw. In which heathen state of mind, I came within view of the house, and stopped to examine it attentively.

It was a **solitary** house, standing in a sadly neglected garden, a pretty even square of some two acres. It was a house of about the time of George the Second; as stiff, as cold, as formal, and in as bad taste, as could possibly be desired by the most loyal admirer of the whole quartet of Georges. It was uninhabited, but had, within a year or two, been cheaply repaired to render it habitable; I say cheaply, because the work had been done in a surface manner, and was already decaying as to the paint and plaster, though the colours were fresh. A lopsided board drooped over the garden wall, announcing that it was "to let on very reasonable terms, well furnished." It was much too closely and heavily shadowed by trees, and, in particular, there were six tall poplars before the front windows, which were excessively melancholy, and the site of which had been extremely ill chosen.

It was easy to see that it was an avoided house -- a house that was shunned by the village, to which my eye was guided by a church spire some half a mile off -- a house that nobody would take. And the natural inference was, that it had the reputation of being a haunted house.

No period within the four-and-twenty hours of day and night is so solemn to me, as the early morning. In the summertime, I often rise very early, and repair to my room to do a day's work before breakfast, and I am always on those occasions deeply impressed by the stillness and solitude around me. Besides that there is something awful in the being surrounded by familiar faces asleep -- in the knowledge that those who are dearest to us and to whom we are dearest, are profoundly unconscious of us, in an impassive state, anticipative of that mysterious condition to which we are all tending -- the stopped life, the broken threads of yesterday, the deserted seat, the closed book, the unfinished but abandoned occupation, all are images of Death. The tranquillity of the hour is the tranquillity of Death. The colour and the chill have the same association. Even a certain air that familiar household objects take upon them when they first emerge from the shadows of the night into the morning, of being newer, and as they used to be long ago, has its counterpart in the subsidence of the worn face of maturity or age, in death, into the old youthful look. Moreover,

Solitary – *Lonely*
Shun – *Avoid*
Solemn – *Sombre*
Tranquil – *Calm*

I once saw the apparition of my father, at this hour. He was alive and well, and nothing ever came of it, but I saw him in the daylight, sitting with his back towards me, on a seat that stood beside my bed. His head was resting on his hand, and whether he was slumbering or grieving, I could not discern. Amazed to see him there, I sat up, moved my position, leaned out of bed, and watched him. As he did not move, I spoke to him more than once. As he did not move then, I became alarmed and laid my hand upon his shoulder, as I thought -- and there was no such thing.

For all these reasons, and for others less easily and briefly statable, I find the early morning to be my most ghostly time. Any house would be more or less haunted, to me, in the early morning; and a haunted house could scarcely address me to greater advantage than then. I walked on into the village, with the desertion of this house upon my mind, and I found the landlord of the little inn, sanding his door step. I bespoke breakfast, and broached the subject of the house.

"Is it haunted?" I asked.

The landlord looked at me, shook his head, and answered, "I say nothing."

"Then it *is* haunted?"

"Well!" cried the landlord, in an outburst of frankness that had the appearance of desperation --"I wouldn't sleep in it."

"Why not?"

"If I wanted to have all the bells in a house ring, with nobody to ring 'em; and all the doors in a house bang, with nobody to bang 'em; and all sorts of feet treading about, with no feet there; why, then," said the landlord, "I'd sleep in that house."

"Is anything seen there?"

The landlord looked at me again, and then, with his former appearance of desperation, called down his stable yard for "Ikey!"

The call produced a high-shouldered young fellow, with a round red face, a short crop of sandy hair, a very broad humoros mouth, a turned-up nose, and a great sleeved waistcoat of purple bars, with mother-of-pearl buttons that seemed to be growing upon him, and to be in a fair way -- if it were not pruned -- of covering his head and overunning his boots.

Apparition – *Ghost*
Discern – *Distinguish*
Haunt – *Trouble, Loiter*
Desertion – *Abandonment*

"This gentleman wants to know," said the landlord, "if anything's seen at the Poplars."

"'Ooded woman with a howl," said Ikey, in a state of great freshness.

"Do you mean a cry?"

"I mean a bird, sir."

"A hooded woman with an owl. Dear me! Did you ever see her?"

"I seen the howl."

"Never the woman?"

"Not so plain as the howl, but they always keeps together."

"Has anybody ever seen the woman as plainly as the owl?"

"Lord bless you, sir! Lots."

"Who?" "Lord bless you, sir! Lots."

"The general dealer opposite, for instance, who is opening his shop?"

"Perkins? Bless you, Perkins wouldn't go a-nigh the place. No!" observed the young man, with considerable feeling; "he an't overwise, an't Perkins, but he an't such a fool as that."

(Here, the landlord murmured his confidence in Perkins's knowing better.)

"Who is -- or who was -- the hooded woman with the owl? Do you know?"

"Well!" said Ikey, holding up his cap with one hand while he scratched his head with the other, "they say, in general, that she was murdered, and the howl he 'ooted the while."

This very concise summary of the facts was all I could learn, except that a young man, as hearty and likely a young man as ever I see, had been took with fits and held down in 'em, after seeing the hooded woman. Also, that a personage, dimly described as "a hold chap, a sort of one-eyed tramp, answering to the name of Joby, unless you challenged him as Greenwood, and then he said, 'Why not? And even if so, mind your own business,'" had encountered the hooded woman, a matter of five or six times. But, I was not materially assisted by these witnesses in as much as the first was in California, and the last was, as Ikey said (and he was confirmed by the landlord), Anywheres.

Now, although I regard with a hushed and solemn fear, the mysteries, between which and this state of existence is interposed the barrier of the great trial and change that fall

Murmur – *Speak softly*
Concise – *Brief*
Personage – *Celebrity*
Interpose – *Interject*

on all the things that live; and although I have not the audacity to pretend that I know anything of them; I can no more reconcile the mere banging of doors, ringing of bells, creaking of boards, and such-like insignificances, with the majestic beauty and pervading analogy of all the Divine rules that I am permitted to understand, than I had been able, a little while before, to yoke the spiritual intercourse of my fellow-traveller to the chariot of the rising sun. Moreover, I had lived in two haunted houses -- both abroad. In one of these, an old Italian palace, which bore the reputation of being very badly haunted indeed, and which had recently been twice abandoned on that account, I lived eight months, most tranquilly and pleasantly, notwithstanding that the house had a score of mysterious bedrooms, which were never used, and possessed, in one large room in which I sat reading, times out of number at all hours, and next to which I slept, a haunted chamber of the first pretensions. I gently hinted these considerations to the landlord. And as to this particular house having a bad name, I reasoned with him, Why, how many things had bad names undeservedly, and how easy it was to give bad names, and did he not think that if he and I were persistently to whisper in the village that any weird-looking old drunken tinker of the neighbourhood had sold himself to the Devil, he would come in time to be suspected of that commercial venture! All this wise talk was perfectly ineffective with the landlord, I am bound to confess, and was as dead a failure as ever I made in my life.

To cut this part of the story short, I was piqued about the haunted house, and was already half resolved to take it. So, after breakfast, I got the keys from Perkins's brother-in-law (a whip and harness maker, who keeps the Post Office, and is under submission to a most rigorous wife of the Doubly Seceding Little Emmanuel persuasion), and went up to the house, attended by my landlord and by Ikey.

Within, I found it, as I had expected, transcendently dismal. The slowly changing shadows waved on it from the heavy trees, were doleful in the last degree; the house was ill-placed, ill-built, ill-planned, and ill-fitted. It was damp, it was not free from dry rot, there was a flavour of rats in it, and it was the gloomy victim of that indescribable decay which settles on all the work of man's hands whenever it's not turned to man's account. The kitchens and offices were too large, and

Reconcile – *Settle*
Abandon – *Dump*
Pretension
– *Pretence*
Piqued – *Annoyed, Offend*

too remote from each other. Above stairs and below, waste tracts of passage intervened between patches of fertility represented by rooms; and there was a mouldy old well with a green growth upon it, hiding like a murderous trap, near the bottom of the back stairs, under the double row of bells. One of these bells was labelled, on a black ground in faded white letters, Master B. This, they told me, was the bell that rang the most.

"Who was Master B.?" I asked. "Is it known what he did while the owl hooted?"

"Rang the bell," said Ikey.

I was rather struck by the prompt dexterity with which this young man pitched his fur cap at the bell, and rang it himself. It was a loud, unpleasant bell, and made a very disagreeable sound. The other bells were inscribed according to the names of the rooms to which their wires were conducted as "Picture Room," "Double Room," "Clock Room," and the like. Following Master B.'s bell to its source I found that young gentleman to have had but indifferent third-class accommodation in a triangular cabin under the cock-loft, with a corner fireplace which Master B. must have been exceedingly small if he were ever able to warm himself at, and a corner chimney-piece like a pyramidal staircase to the ceiling for Tom Thumb. The papering of one side of the room had dropped down bodily, with fragments of plaster adhering to it, and almost blocked up the door. It appeared that Master B., in his spiritual condition, always made a point of pulling the paper down. Neither the landlord nor Ikey could suggest why he made such a fool of himself.

Except that the house had an immensely large rambling loft at top, I made no other discoveries. It was moderately well furnished, but sparely. Some of the furniture -- say, a third -- was as old as the house; the rest was of various periods within the last half century. I was referred to a corn-chandler in the market place of the county town to treat for the house. I went that day, and I took it for six months.

It was just the middle of October when I moved in with my maiden sister (I venture to call her eight-and-thirty, she is so very handsome, sensible, and engaging). We took with us, a deaf stable-man, my bloodhound Turk, two women servants, and a young person called an Odd Girl. I have reason

Dexterity – *Deftness, Agility*
immense – *Enormous*
Rambling – *Confused*

to record of the attendant last enumerated, who was one of the Saint Lawrence's Union Female Orphans, that she was a fatal mistake and a disastrous engagement.

The year was dying early, the leaves were falling fast, it was a raw cold day when we took possession, and the gloom of the house was most depressing. The cook (an amiable woman, but of a weak turn of intellect) burst into tears on beholding the kitchen, and requested that her silver watch might be delivered over to her sister (2 Tuppintock's Gardens, Liggs's Walk, Clapham Rise), in the event of anything happening to her from the damp. Streaker, the housemaid, feigned cheerfulness, but was the greater martyr. The Odd Girl, who had never been in the country, alone was pleased, and made arrangements for sowing an acorn in the garden outside the scullery window, and rearing an oak.

We went, before dark, through all the natural -- as opposed to supernatural -- miseries incidental to our state. Dispiriting reports ascended (like the smoke) from the basement in volumes, and descended from the upper rooms. There was no rolling-pin, there was no salamander (which failed to surprise me, for I don't know what it is), there was nothing in the house, what there was, was broken, the last people must have lived like pigs, what could the meaning of the landlord be? Through these distresses, the Odd Girl was cheerful and exemplary. But within four hours after dark we had got into a supernatural groove, and the Odd Girl had seen "Eyes," and was in hysterics.

My sister and I had agreed to keep the haunting strictly to ourselves, and my impression was, and still is, that I had not left Ikey, when he helped to unload the cart, alone with the women, or any one of them, for one minute. Nevertheless, as I say, the Odd Girl had "seen Eyes" (no other explanation could ever be drawn from her), before nine, and by ten o'clock had had as much vinegar applied to her as would pickle a handsome salmon.

I leave a discerning public to judge of my feelings, when, under these untoward circumstances, at about half-past ten o'clock Master B.'s bell began to ring in a most infuriated manner, and Turk howled until the house resounded with his lamentations!

I hope I may never again be in a state of mind so unchristian as the mental frame in which I lived for some weeks,

Enumerate – *To ascertain, Recount*
Fatal – *Deadly*
Feign – *Pretend*
Exemplary – *Ideal, Commendable*

respecting the memory of Master B. Whether his bell was rung by rats, or mice, or bats, or wind, or what other accidental vibration, or sometimes by one cause, sometimes another, and sometimes by collusion, I don't know; but, certain it is, that it did ring two nights out of three, until I conceived the happy idea of twisting Master B.'s neck -- in other words, breaking his bell short off -- and silencing that young gentleman, as to my experience and belief, forever.

But, by that time, the Odd Girl had developed such improving powers of catalepsy, that she had become a shining example of that very inconvenient disorder. She would stiffen, like a Guy Fawkes endowed with unreason, on the most irrelevant occasions. I would address the servants in a lucid manner, pointing out to them that I had painted Master B.'s room and balked the paper, and taken Master B.'s bell away and balked the ringing, and if they could suppose that that confounded boy had lived and died, to clothe himself with no better behaviour than would most unquestionably have brought him and the sharpest particles of a birch-broom into close acquaintance in the present imperfect state of existence, could they also suppose a mere poor human being, such as I was, capable by those contemptible means of counteracting and limiting the powers of the disembodied spirits of the dead, or of any spirits? -- I say I would become emphatic and cogent, not to say rather complacent, in such an address, when it would all go for nothing by reason of the Odd Girl's suddenly stiffening from the toes upward, and glaring among us like a parochial petrifaction.

Streaker, the housemaid, too, had an attribute of a most discomfiting nature. I am unable to say whether she was of an usually lymphatic temperament, or what else was the matter with her, but this young woman became a mere distillery for the production of the largest and most transparent tears I ever met with. Combined with these characteristics, was a peculiar tenacity of hold in those specimens, so that they didn't fall, but hung upon her face and nose. In this condition, and mildly and deplorably shaking her head, her silence would throw me more heavily than the Admirable Crichton could have done in a verbal disputation for a purse of money. Cook, likewise, always covered me with confusion as with a garment, by neatly winding up the session with the protest that the Ouse was wearing

Collusion – Conspiracy
Confound – Stun, Surprise
Emphatic – Forceful
Tenacity – Stubbornness

her out, and by meekly repeating her last wishes regarding her silver watch.

As to our nightly life, the contagion of suspicion and fear was among us, and there is no such contagion under the sky. Hooded woman? According to the accounts, we were in a perfect Convent of hooded women. Noises? With that contagion downstairs, I myself have sat in the dismal parlour, listening, until I have heard so many and such strange noises, that they would have chilled my blood if I had not warmed it by dashing out to make discoveries. Try this in bed, in the dead of the night, try this at your own comfortable fireside, in the life of the night. You can fill any house with noises, if you will, until you have a noise for every nerve in your nervous system.

I repeat; the contagion of suspicion and fear was among us, and there is no such contagion under the sky. The women (their noses in a chronic state of excoriation from smelling-salts) were always primed and loaded for a swoon, and ready to go off with hair-triggers. The two elder detached the Odd Girl on all expeditions that were considered doubly hazardous, and she always established the reputation of such adventures by coming back cataleptic. If the Cook or Streaker went overhead after dark, we knew we should presently hear a bump on the ceiling; and this took place so constantly, that it was as if a fighting man were engaged to go about the house, administering a touch of his art which I believe is called The Auctioneer, to every domestic he met with.

It was in vain to do anything. It was in vain to be frightened, for the moment in one's own person, by a real owl, and then to show the owl. It was in vain to discover, by striking an accidental discord on the piano, that Turk always howled at particular notes and combinations. It was in vain to be a Rhadamanthus with the bells, and if an unfortunate bell rang without leave, to have it down inexorably and silence it. It was in vain to fire up chimneys, let torches down the well, charge furiously into suspected rooms and recesses. We changed servants, and it was no better. The new set ran away, and a third set came, and it was no better. At last, our comfortable housekeeping got to be so disorganised and wretched, that I one night dejectedly said to my sister, "Patty, I begin to despair of our getting people to go on with us here, and I think we must give this up."

Meek – *Humble*
Contagion – *Infection*
Dismal – *Miserable*
Vain – *Very Proud*

My sister, who is a woman of immense spirit, replied, "No, John, don't give it up. Don't be beaten, John. There is another way."

"And what is that?" said I.

"John," returned my sister, "if we are not to be driven out of this house, and that for no reason whatever, that is apparent to you or me, we must help ourselves and take the house wholly and solely into our own hands."

"But, the servants," said I.

"Have no servants," said my sister, boldly.

Like most people in my grade of life, I had never thought of the possibility of going on without those faithful obstructions. The notion was so new to me when suggested, that I looked very doubtful. "We know they come here to be frightened and infect one another, and we know they are frightened and do infect one another," said my sister.

"With the exception of Bottles," I observed, in a meditative tone.

(The deaf stable-man. I kept him in my service, and still keep him, as a phenomenon of moroseness not to be matched in England.)

"To be sure, John," assented my sister; "except Bottles. And what does that go to prove? Bottles talks to nobody, and hears nobody unless he is absolutely roared at, and what alarm has Bottles ever given, or taken! None."

This was perfectly true; the individual in question having retired, every night at ten o'clock, to his bed over the coach-house, with no other company than a pitchfork and a pail of water. That the pail of water would have been over me, and the pitchfork through me, if I had put myself without announcement in Bottles's way after that minute, I had deposited in my own mind as a fact worth remembering. Neither had Bottles ever taken the least notice of any of our many uproars. An imperturbable and speechless man, he had sat at his supper, with Streaker present in a swoon, and the Odd Girl marble, and had only put another potato in his cheek, or profited by the general misery to help himself to beefsteak pie.

"And so," continued my sister, "I exempt Bottles. And considering, John, that the house is too large, and perhaps too lonely, to be kept well in hand by Bottles, you, and me, I propose that we cast about among our friends for a certain selected

Immense
– Enormous, Huge
Notion – Concept
Morose – Depressed
Assent – Concur, To agree

number of the most reliable and willing -- form a Society here for three months -- wait upon ourselves and one another -- live cheerfully and socially -- and see what happens."

I was so charmed with my sister, that I embraced her on the spot, and went into her plan with the greatest ardour.

We were then in the third week of November; but, we took our measures so vigorously, and were so well seconded by the friends in whom we confided, that there was still a week of the month unexpired when our party all came down together merrily, and mustered in the haunted house.

I will mention, in this place, two small changes that I made while my sister and I were yet alone. It occurring to me as not improbable that Turk howled in the house at night, partly because he wanted to get out of it, I stationed him in his kennel outside, but unchained; and I seriously warned the village that any man who came in his way must not expect to leave him without a rip in his own throat. I then casually asked Ikey if he were a judge of a gun? On his saying, "Yes, sir, I knows a good gun when I sees her," I begged the favour of his stepping up to the house and looking at mine.

"She's a true one, sir," said Ikey, after inspecting a double-barrelled rifle that I bought in New York a few years ago. "No mistake about her, sir."

"Ikey," said I, "don't mention it; I have seen something in this house."

"No, sir?" he whispered, greedily opening his eyes. "'Ooded lady, sir?"

"Don't be frightened," said I. "It was a figure rather like you."

"Lord, sir?"

"Ikey!" said I, shaking hands with him warmly. I may say affectionately; "if there is any truth in these ghost stories, the greatest service I can do you, is, to fire at that figure. And I promise you, by Heaven and earth, I will do it with this gun if I see it again!"

The young man thanked me, and took his leave with some little precipitation, after declining a glass of liquor. I imparted my secret to him, because I had never quite forgotten his throwing his cap at the bell; because I had, on another occasion, noticed something very like a fur cap, lying not far from the bell, one night when it had burst out ringing; and because

Embrace – *Hug*
Ardour – *Passion, Tervous*
Improbable – *Unlikely*
Decline – *Refuse*

I had remarked that we were at our ghostliest whenever he came up in the evening to comfort the servants. Let me do Ikey no injustice. He was afraid of the house, and believed in its being haunted; and yet he would play false on the haunting side, so surely as he got an opportunity. The Odd Girl's case was exactly similar. She went about the house in a state of real terror, and yet lied monstrously and wilfully, and invented many of the alarms she spread, and made many of the sounds we heard. I had had my eye on the two, and I know it. It is not necessary for me, here, to account for this preposterous state of mind; I content myself with remarking that it is familiarly known to every intelligent man who has had fair medical, legal, or other watchful experience; that it is as well established and as common a state of mind as any with which observers are acquainted; and that it is one of the first elements, above all others, rationally to be suspected in, and strictly looked for, and separated from, any question of this kind.

To return to our party. The first thing we did when we were all assembled, was, to draw lots for bedrooms. That done, and every bedroom, and, indeed, the whole house, having been minutely examined by the whole body, we allotted the various household duties, as if we had been on a gipsy party, or a yachting party, or a hunting party, or were ship-wrecked. I then recounted the floating rumours concerning the hooded lady, the owl, and Master B. with others, still more filmy, which had floated about during our occupation, relative to some ridiculous old ghost of the female gender who went up and down, carrying the ghost of a round table; and also to an impalpable Jackass, whom nobody was ever able to catch. Some of these ideas I really believe our people below had communicated to one another in some diseased way, without conveying them in words. We then gravely called one another to witness, that we were not there to be deceived, or to deceive -- which we considered pretty much the same thing -- and that, with a serious sense of responsibility, we would be strictly true to one another, and would strictly follow out the truth. The understanding was established, that anyone who heard unusual noises in the night, and who wished to trace them, should knock at my door; lastly, that on twelfth night, the last night of holy Christmas, all our individual experiences since that then present hour of our coming together in the haunted house, should be brought to light for the good of all;

Terror – *Horror*
Preposterous – *Ridiculous*
Acquaint – *To make familiar, Aware*
Deceive – *Mislead*

and that we would hold our peace on the subject till then, unless on some remarkable provocation to break silence.

We were, in number and in character, as follows:

First -- to get my sister and myself out of the way -- there were we two. In the drawing of lots, my sister drew her own room, and I drew Master B.'s. Next, there was our first cousin John Herschel, so called after the great astronomer, than whom I suppose a better man at a telescope does not breathe. With him, was his wife, a charming creature to whom he had been married in the previous spring. I thought it (under the circumstances) rather imprudent to bring her, because there is no knowing what even a false alarm may do at such a time; but I suppose he knew his own business best, and I must say that if she had been my wife, I never could have left her endearing and bright face behind. They drew the clock room. Alfred Starling, an uncommonly agreeable young fellow of eight-and-twenty for whom I have the greatest liking, was in the double room; mine, usually, and designated by that name from having a dressing room within it, with two large and cumbersome windows, which no wedges I was ever able to make, would keep from shaking, in any weather, wind or no wind. Alfred is a young fellow who pretends to be "fast" (another word for loose, as I understand the term), but who is much too good and sensible for that nonsense, and who would have distinguished himself before now, if his father had not unfortunately left him a small independence of two hundred a year, on the strength of which his only occupation in life has been to spend six. I am in hopes, however, that his banker may break, or that he may enter into some speculation guaranteed to pay twenty per cent; for, I am convinced that if he could only be ruined, his fortune is made. Belinda Bates, bosom friend of my sister, and a most intellectual, amiable, and delightful girl, got the picture room. She has a fine genius for poetry, combined with real business earnestness, and "goes in" -- to use an expression of Alfred's -- for woman's mission, woman's rights, woman's wrongs, and everything that is woman's with a capital W, or is not and ought to be, or is and ought not to be. "Most praiseworthy, my dear, and Heaven prosper you!" I whispered to her on the first night of my taking leave of her at the picture room door, "but don't overdo it. And in respect of the great necessity there is, my darling, for more employments being

Provocation – *Aggravation*
Cumbersome – *Unwieldy, Troublesome*
Amiable – *Friendly*
Earnest – *Serious, True, Sincere*

within the reach of woman than our civilisation has as yet assigned to her, don't fly at the unfortunate men, even those men who are at first sight in your way, as if they were the natural oppressors of your sex; for, trust me, Belinda, they do sometimes spend their wages among wives and daughters, sisters, mothers, aunts, and grandmothers; and the play is, really, not all Wolf and Red Riding Hood, but has other parts in it." However, I digress.

Belinda, as I have mentioned, occupied the picture room. We had but three other chambers: the Corner Room, the cupboard room, and the garden room. My old friend, Jack Governor, "slung his hammock," as he called it, in the corner room. I have always regarded Jack as the finest-looking sailor that ever sailed. He is gray now, but as handsome as he was a quarter of a century ago -- nay, handsomer. A portly, cheery, well-built figure of a broad-shouldered man, with a frank smile, a brilliant dark eye, and a rich dark eyebrow. I remember those under darker hair, and they look all the better for their silver setting. He has been wherever his Union namesake flies, has Jack, and I have met old shipmates of his, away in the Mediterranean and on the other side of the Atlantic, who have beamed and brightened at the casual mention of his name, and have cried, "You know Jack Governor? Then you know a prince of men!" That he is! And so unmistakably a naval officer, that if you were to meet him coming out of an Esquimaux snow-hut in seal's skin, you would be vaguely persuaded he was in full naval uniform.

Jack once had that bright clear eye of his on my sister; but, it fell out that he married another lady and took her to South America, where she died. This was a dozen years ago or more. He brought down with him to our haunted house a little cask of salt beef; for, he is always convinced that all salt beef not of his own pickling, is mere carrion, and invariably, when he goes to London, packs a piece in his portmanteau. He had also volunteered to bring with him one "Nat Beaver," an old comrade of his, captain of a merchantman. Mr. Beaver, with a thick-set wooden face and figure, and apparently as hard as a block all over, proved to be an intelligent man, with a world of watery experiences in him, and great practical knowledge. At times, there was a curious nervousness about him, apparently the lingering result of some old illness; but, it seldom lasted many minutes. He got

Oppressor – *Tormenter*
Digress – *Deviate*
Carrion – *Rotten, Dead and Decaying*

the cupboard room, and lay there next to Mr. Undery, my friend and solicitor, who came down, in an amateur capacity, "to go through with it," as he said, and who plays whist better than the whole Law List, from the red cover at the beginning to the red cover at the end.

I never was happier in my life, and I believe it was the universal feeling among us. Jack Governor, always a man of wonderful resources, was Chief Cook, and made some of the best dishes I ever ate, including unapproachable curries. My sister was pastrycook and confectioner. Starling and I were Cook's Mate, turn and turn about, and on special occasions the chief cook "pressed" Mr. Beaver. We had a great deal of outdoor sport and exercise, but nothing was neglected within, and there was no ill-humour or misunderstanding among us, and our evenings were so delightful that we had at least one good reason for being reluctant to go to bed.

We had a few night alarms in the beginning. On the first night, I was knocked up by Jack with a most wonderful ship's lantern in his hand, like the gills of some monster of the deep, who informed me that he "was going aloft to the main truck," to have the weathercock down. It was a stormy night and I remonstrated; but Jack called my attention to its making a sound like a cry of despair, and said somebody would be "hailing a ghost" presently, if it wasn't done. So, up to the top of the house, where I could hardly stand for the wind, we went, accompanied by Mr. Beaver; and there Jack, lantern and all, with Mr. Beaver after him, swarmed up to the top of a cupola, some two dozen feet above the chimneys, and stood upon nothing particular, coolly knocking the weathercock off, until they both got into such good spirits with the wind and the height, that I thought they would never come down. Another night, they turned out again, and had a chimney-cowl off. Another night, they cut a sobbing and gulping water pipe away. Another night, they found out something else. On several occasions, they both, in the coolest manner, simultaneously dropped out of their respective bedroom windows, hand over hand by their counterpanes, to "overhaul" something mysterious in the garden.

Solicitor – *Lawyer*
Reluctant
– *Unwilling*
Aloft – *Uphill*
Overhaul – *Repair*

The engagement among us was faithfully kept, and nobody revealed anything. All we knew was, if anyone's room were haunted, no one looked the worse for it.

CHAPTER II - THE GHOST IN MASTER B'S ROOM

When I established myself in the triangular garret which had gained so distinguished a reputation, my thoughts naturally turned to Master B. My speculations about him were uneasy and manifold. Whether his Christian name was Benjamin, Bissextile (from his having been born in Leap Year), Bartholomew, or Bill. Whether the initial letter belonged to his family name, and that was Baxter, Black, Brown, Barker, Buggins, Baker, or Bird. Whether he was a foundling, and had been baptized B. Whether he was a lion-hearted boy, and B. was short for Briton, or for Bull. Whether he could possibly have been kith and kin to an illustrious lady who brightened my own childhood, and had come of the blood of the brilliant Mother Bunch?

With these profitless meditations I tormented myself much. I also carried the mysterious letter into the appearance and pursuits of the deceased; wondering whether he dressed in Blue, wore Boots (he couldn't have been Bald), was a boy of Brains, liked Books, was good at Bowling, had any skill as a Boxer, even in his Buoyant Boyhood Bathed from a Bathing-machine at Bognor, Bangor, Bournemouth, Brighton, or Broadstairs, like a Bounding Billiard Ball?

So, from the first, I was haunted by the letter B.

It was not long before I remarked that I never by any hazard had a dream of Master B., or of anything belonging to him. But, the instant I awoke from sleep, at whatever hour of the night, my thoughts took him up, and roamed away, trying to attach his initial letter to something that would fit it and keep it quiet.

For six nights, I had been worried this in Master B.'s room, when I began to perceive that things were going wrong.

The first appearance that presented itself was early in the morning when it was but just daylight and no more. I was standing shaving at my glass, when I suddenly discovered, to my consternation and amazement, that I was shaving -- not myself -- I am fifty -- but a boy. Apparently Master B.!

I trembled and looked over my shoulder; nothing there. I looked again in the glass, and distinctly saw the features and expression of a boy, who was shaving, not to get rid of a beard, but to get one. Extremely troubled in my mind, I took a few turns in the room, and went back to the looking glass,

Garret – *Loft*
Manifold – *Various*
Hazard – *Danger*
Consternation – *Dismay*

resolved to steady my hand and complete the operation in which I had been disturbed. Opening my eyes, which I had shut while recovering my firmness, I now met in the glass, looking straight at me, the eyes of a young man of four or five and twenty. Terrified by this new ghost, I closed my eyes, and made a strong effort to recover myself. Opening them again, I saw, shaving his cheek in the glass, my father, who has long been dead. Nay, I even saw my grandfather too, whom I never did see in my life.

Although naturally much affected by these remarkable visitations, I determined to keep my secret, until the time agreed upon for the present general disclosure. Agitated by a multitude of curious thoughts, I retired to my room, that night, prepared to encounter some new experience of a spectral character. Nor was my preparation needless, for, waking from an uneasy sleep at exactly two o'clock in the morning, what were my feelings to find that I was sharing my bed with the skeleton of Master B.!

I sprang up, and the skeleton sprang up also. I then heard a plaintive voice saying, "Where am I? What is become of me?" and, looking hard in that direction, perceived the ghost of Master B. The young spectre was dressed in an obsolete fashion, or rather, was not so much dressed as put into a case of inferior pepper-and- salt cloth, made horrible by means of shining buttons. I observed that these buttons went, in a double row, over each shoulder of the young ghost, and appeared to descend his back. He wore a frill round his neck. His right hand (which I distinctly noticed to be inky) was laid upon his stomach; connecting this action with some feeble pimples on his countenance, and his general air of nausea, I concluded this ghost to be the ghost of a boy who had habitually taken a great deal too much medicine.

"Where am I?" said the little spectre, in a pathetic voice. "And why was I born in the Calomel days, and why did I have all that Calomel given me?"

I replied, with sincere earnestness, that upon my soul I couldn't tell him.

"Where is my little sister," said the ghost, "and where my angelic little wife, and where is the boy I went to school with?"

I entreated the phantom to be comforted, and above all things to take heart respecting the loss of the boy he went to

Disclosure
– *Revelation*
Spectre – *Ghost*
Feeble – *Weak*
Entreat – *Plead*

school with. I represented to him that probably that boy never did, within human experience, come out well, when discovered. I urged that I myself had, in later life, turned up several boys whom I went to school with, and none of them had at all answered. I expressed my humble belief that that boy never did answer. I represented that he was a mythic character, a delusion, and a snare. I recounted how, the last time I found him, I found him at a dinner party behind a wall of white cravat, with an inconclusive opinion on every possible subject, and a power of silent boredom absolutely Titanic. I related how, on the strength of our having been together at "Old Doylance's," he had asked himself to breakfast with me (a social offence of the largest magnitude); how, fanning my weak embers of belief in Doylance's boys, I had let him in; and how, he had proved to be a fearful wanderer about the earth, pursuing the race of Adam with inexplicable notions concerning the currency, and with a proposition that the Bank of England should, on pain of being abolished, instantly strike off and circulate, God knows how many thousand millions of ten-and-six penny notes.

The ghost heard me in silence, and with a fixed stare. "Barber!" It apostrophised me when I had finished.

"Barber?" I repeated -- for I am not of that profession.

"Condemned," said the ghost, "to shave a constant change of customers -- now, me -- now, a young man -- now, thyself as thou art -- now, thy father -- now, thy grandfather; condemned, too, to lie down with a skeleton every night, and to rise with it every morning --"

(I shuddered on hearing this dismal announcement.)

"Barber! Pursue me!"

I had felt, even before the words were uttered, that I was under a spell to pursue the phantom. I immediately did so, and was in Master B.'s room no longer.

Most people know what long and fatiguing night journeys had been forced upon the witches who used to confess, and who, no doubt, told the exact truth -- particularly as they were always assisted with leading questions, and the torture was always ready. I asseverate that, during my occupation of Master B.'s room, I was taken by the ghost that haunted it, on expeditions fully as long and wild as any of those. Assuredly, I was presented to no shabby old man with a goat's horns and

Humble – *Modest*
Delusion – *Illusion*
Dismal – *Miserable*
Pursue – *Follow*

tail (something between Pan and an old clothesman), holding conventional receptions, as stupid as those of real life and less decent; but, I came upon other things which appeared to me to have more meaning.

Confident that I speak the truth and shall be believed, I declare without hesitation that I followed the ghost, in the first instance on a broom stick, and afterwards on a rocking horse. The very smell of the animal's paint -- especially when I brought it out, by making him warm -- I am ready to swear to. I followed the ghost, afterwards, in a hackney coach; an institution with the peculiar smell of which, the present generation is unacquainted, but to which I am again ready to swear as a combination of stable, dog with the mange, and very old bellows. (In this, I appeal to previous generations to confirm or refute me.) I pursued the phantom, on a headless donkey, at least, upon a donkey who was so interested in the state of his stomach that his head was always down there, investigating it; on ponies, expressly born to kick up behind; on roundabouts and swings, from fairs; in the first cab -- another forgotten institution where the fare regularly got into bed, and was tucked up with the driver.

Not to trouble you with a detailed account of all my travels in pursuit of the ghost of Master B., which were longer and more wonderful than those of Sinbad the Sailor, I will confine myself to one experience from which you may judge of many.

I was marvellously changed. I was myself, yet not myself. I was conscious of something within me, which has been the same all through my life, and which I have always recognised under all its phases and varieties as never altering, and yet I was not the I who had gone to bed in Master B.'s room. I had the smoothest of faces and the shortest of legs, and I had taken another creature like myself, also with the smoothest of faces and the shortest of legs, behind a door, and was confiding to him a proposition of the most astounding nature.

This proposition was, that we should have a Seraglio.

The other creature assented warmly. He had no notion of respectability, neither had I. It was the custom of the East, it was the way of the good Caliph Haroun Alraschid (let me have the corrupted name again for once, it is so scented with sweet memories!), the usage was highly laudable, and most

Conventional –
Conservative
Refute – *Disprove, Rebut*
Astound – *Amaze*
Laudable
– *Creditable*

worthy of imitation. "O, yes! Let us," said the other creature with a jump, "have a Seraglio."

It was not because we entertained the faintest doubts of the meritorious character of the Oriental establishment we proposed to import, that we perceived it must be kept a secret from Miss Griffin. It was because we knew Miss Griffin to be bereft of human sympathies, and incapable of appreciating the greatness of the great Haroun. Mystery impenetrably shrouded from Miss Griffin then, let us entrust it to Miss Bule.

We were ten in Miss Griffin's establishment by Hampstead Ponds; eight ladies and two gentlemen. Miss Bule, whom I judge to have attained the ripe age of eight or nine, took the lead in society. I opened the subject to her in the course of the day, and proposed that she should become the favourite.

Miss Bule, after struggling with the diffidence so natural to, and charming in, her adorable sex, expressed herself as flattered by the idea, but wished to know how it was proposed to provide for Miss Pipson? Miss Bule -- who was understood to have vowed towards that young lady, a friendship, halves, and no secrets, until death, on the Church Service and Lessons complete in two volumes with case and lock -- Miss Bule said she could not, as the friend of Pipson, disguise from herself, or me, that Pipson was not one of the common.

Now, Miss Pipson, having curly hair and blue eyes (which was my idea of anything mortal and feminine that was called fair), I promptly replied that I regarded Miss Pipson in the light of a Fair Circassian.

"And what then?" Miss Bule pensively asked.

I replied that she must be inveigled by a merchant, brought to me veiled, and purchased as a slave. The other creature had already fallen into the second male place in the State, and was set apart for Grand Vizier. He afterwards resisted this disposal of events, but had his hair pulled until he yielded.

"Shall I not be jealous?" Miss Bule inquired, casting down her eyes.

"Zobeide, no," I replied; "you will ever be the favourite Sultana; the first place in my heart, and on my throne, will be ever yours."

Miss Bule, upon that assurance, consented to propound the idea to her seven beautiful companions. It occurring to

Meritorious – *Commendable*
Diffidence – *Shyness*
Pensive – *Thoughtful*
Inveigle – *To lure, Induce*

me, in the course of the same day, that we knew we could trust a grinning and good-natured soul called Tabby, who was the serving drudge of the house, and had no more figure than one of the beds, and upon whose face there was always more or less black-lead, I slipped into Miss Bule's hand after supper, a little note to that effect; dwelling on the black-lead as being in a manner deposited by the finger of Providence, pointing Tabby out for Mesrour, the celebrated chief of the Blacks of the Hareem.

There were difficulties in the formation of the desired institution, as there are in all combinations. The other creature showed himself of a low character, and, when defeated in aspiring to the throne, pretended to have conscientious scruples about prostrating himself before the Caliph; wouldn't call him Commander of the Faithful; spoke of him slightingly and inconsistently as a mere "chap"; said he, the other creature, "wouldn't play" -- Play! -- and was otherwise coarse and offensive. This meanness of disposition was, however, put down by the general indignation of an united Seraglio, and I became blessed in the smiles of eight of the fairest of the daughters of men.

The smiles could only be bestowed when Miss Griffin was looking another way, and only then in a very wary manner, for there was a legend among the followers of the Prophet that she saw with a little round ornament in the middle of the pattern on the back of her shawl. But everyday after dinner, for an hour, we were all together, and then the Favourite and the rest of the Royal Hareem competed who should most beguile the leisure of the Serene Haroun reposing from the cares of State -- which were generally, as in most affairs of State, of an arithmetical character, the Commander of the Faithful being a fearful boggler at a sum.

On these occasions, the devoted Mesrour, chief of the Blacks of the Hareem, was always in attendance (Miss Griffin usually ringing for that officer, at the same time, with great vehemence), but never acquitted himself in a manner worthy of his historical reputation. In the first place, his bringing a broom into the Divan of the Caliph, even when Haroun wore on his shoulders the red robe of anger (Miss Pipson's pelisse), though it might be got over for the moment, was never to be quite satisfactorily accounted for. In the second place, his breaking out into grinning exclamations of "Lork

Drudge – *Worker*
Conscientious – *Careful*
Beguile – *Entice*
Vehemence – *Forcefulness*

you pretties!" was neither Eastern nor respectful. In the third place, when specially instructed to say "Bismillah!" he always said "Hallelujah!" This officer, unlike his class, was too good-humoured altogether, kept his mouth open far too wide, expressed approbation to an incongruous extent, and even once -- it was on the occasion of the purchase of the Fair Circassian for five hundred thousand purses of gold, and cheap, too -- embraced the Slave, the Favourite, and the Caliph, all round. (Parenthetically let me say God bless Mesrour, and may there have been sons and daughters on that tender bosom, softening many a hard day since!)

Miss Griffin was a model of propriety, and I am at a loss to imagine what the feelings of the virtuous woman would have been, if she had known, when she paraded us down the Hampstead Road two and two, that she was walking with a stately step at the head of Polygamy and Mahomedanism. I believe that a mysterious and terrible joy with which the contemplation of Miss Griffin, in this unconscious state, inspired us, and a grim sense prevalent among us that there was a dreadful power in our knowledge of what Miss Griffin (who knew all things that could be learnt out of book) didn't know, were the main-spring of the preservation of our secret. It was wonderfully kept, but was once upon the verge of self-betrayal. The danger and escape occurred upon a Sunday. We were all ten ranged in a conspicuous part of the gallery at church, with Miss Griffin at our head -- as we were every Sunday -- advertising the establishment in an unsecular sort of way -- when the description of Solomon in his domestic glory happened to be read. The moment that monarch was thus referred to, conscience whispered me, "Thou, too, Haroun!" The officiating minister had a cast in his eye, and it assisted conscience by giving him the appearance of reading personally at me. A crimson blush, attended by a fearful perspiration, suffused my features. The Grand Vizier became more dead than alive, and the whole Seraglio reddened as if the sunset of Bagdad shone direct upon their lovely faces. At this portentous time the awful Griffin rose, and balefully surveyed the children of Islam. My own impression was, that Church and State had entered into a conspiracy with Miss Griffin to expose us, and that we should all be put into white sheets, and exhibited in the centre aisle. But, so Westerly -- if I may be allowed the expression as opposite to Eastern associations --

Incongruous – *Out of place*
Grim – *Bleak*
Verge – *Border*
Portentous – *Pompous, Self-important*

was Miss Griffin's sense of rectitude, that she merely suspected apples, and we were saved.

I have called the Seraglio, united. Upon the question, solely, whether the Commander of the Faithful durst exercise a right of kissing in that sanctuary of the palace, were its peerless inmates divided. Zobeide asserted a counter-right in the Favourite to scratch, and the fair Circassian put her face, for refuge, into a green baize bag, originally designed for books. On the other hand, a young antelope of transcendent beauty from the fruitful plains of Camden Town (whence she had been brought, by traders, in the half-yearly caravan that crossed the intermediate desert after the holidays), held more liberal opinions, but stipulated for limiting the benefit of them to that dog, and son of a dog, the Grand Vizier--who had no rights, and was not in question. At length, the difficulty was compromised by the installation of a very youthful slave as Deputy. She, raised upon a stool, officially received upon her cheeks the salutes intended by the gracious Haroun for other Sultanas, and was privately rewarded from the coffers of the Ladies of the Hareem.

And now it was, at the full height of enjoyment of my bliss, that I became heavily troubled. I began to think of my mother, and what she would say to my taking home at Midsummer eight of the most beautiful of the daughters of men, but all unexpected. I thought of the number of beds we made up at our house, of my father's income, and of the baker, and my despondency redoubled. The Seraglio and malicious Vizier, divining the cause of their Lord's unhappiness, did their utmost to augment it. They professed unbounded fidelity, and declared that they would live and die with him. Reduced to the utmost wretchedness by these protestations of attachment, I lay awake, for hours at a time, ruminating on my frightful lot. In my despair, I think I might have taken an early opportunity of falling on my knees before Miss Griffin, avowing my resemblance to Solomon, and praying to be dealt with according to the outraged laws of my country, if an unthought-of means of escape had not opened before me.

One day, we were out walking, two and two -- on which occasion the Vizier had his usual instructions to take note of the boy at the turn-pike, and if he profanely gazed (which he always did) at the beauties of the Hareem, to have him

Rectitude –
Righteousness
Refuge – *Sanctuary*
Augment
– *Supplement*
Profanely
– *Contrarily*

bowstrung in the course of the night -- and it happened that our hearts were veiled in gloom. An unaccountable action on the part of the antelope had plunged the State into disgrace. That charmer, on the representation that the previous day was her birthday, and that vast treasures had been sent in a hamper for its celebration (both baseless assertions), had secretly but most pressingly invited thirty-five neighbouring princes and princesses to a ball and supper, with a special stipulation that they were "not to be fetched till twelve." This wandering of the antelope's fancy, led to the surprising ar-rival at Miss Griffin's door, in divers equipages and under various escorts, of a great company in full dress, who were deposited on the top step in a flush of high expectancy, and who were dismissed in tears. At the beginning of the dou-ble knocks attendant on these ceremonies, the antelope had retired to a back attic, and bolted herself in; and at every new arrival, Miss Griffin had gone so much more and more distracted, that at last she had been seen to tear her front. Ultimate capitulation on the part of the offender, had been followed by solitude in the linen-closet, bread and water and a lecture to all, of vindictive length, in which Miss Griffin had used expressions: Firstly, "I believe you all of you knew of it;" Secondly, "Every one of you is as wicked as another;" Thirdly, "A pack of little wretches."

Under these circumstances, we were walking drearily along; and I especially, with my Moosulmaun responsibili-ties heavy on me, was in a very low state of mind; when a strange man accosted Miss Griffin, and, after walking on at her side for a little while and talking with her, looked at me. Supposing him to be a minion of the law, and that my hour was come, I instantly ran away, with the general purpose of making for Egypt.

The whole Seraglio cried out, when they saw me making off as fast as my legs would carry me (I had an impression that the first turning on the left, and round by the public house, would be the shortest way to the Pyramids), Miss Griffin screamed after me, the faithless Vizier ran after me, and the boy at the turn-pike dodged me into a corner, like a sheep, and cut me off. Nobody scolded me when I was taken and brought back; Miss Griffin only said, with a stunning gentleness, "This was very curious! Why had I run away when the gentleman looked at me?"

Vindictive – *Spiteful*
Accost – *Approach*
Minion – *Follower*
Dodge – *To Move aside, evade*

If I had had any breath to answer with, I dare say I should have made no answer; having no breath, I certainly made none. Miss Griffin and the strange man took me between them, and walked me back to the palace in a sort of state; but not at all (as I couldn't help feeling, with astonishment) in culprit state.

When we got there, we went into a room by ourselves, and Miss Griffin called in to her assistance, Mesrour, chief of the dusky guards of the Hareem. Mesrour, on being whispered to, began to shed tears. "Bless you, my precious!" said that officer, turning to me; "your Pa's took bitter bad!"

I asked, with a fluttered heart, "Is he very ill?"

"Lord temper the wind to you, my lamb!" said the good Mesrour, kneeling down, that I might have a comforting shoulder for my head to rest on, "your Pa's dead!"

Haroun Alraschid took to flight at the words; the Seraglio vanished; from that moment, I never again saw one of the eight of the fairest of the daughters of men.

I was taken home, and there was debt at home as well as death, and we had a sale there. My own little bed was so superciliously looked upon by a power unknown to me, hazily called "The Trade," that a brass coal-scuttle, a roasting-jack, and a birdcage, were obliged to be put into it to make a lot of it, and then it went for a song. So I heard mentioned, and I wondered what song, and thought what a dismal song it must have been to sing!

Then, I was sent to a great, cold, bare, school of big boys; where everything to eat and wear was thick and clumpy, without being enough; where everybody, largo and small, was cruel; where the boys knew all about the sale, before I got there, and asked me what I had fetched, and who had bought me, and hooted at me, "Going, going, gone!" I never whispered in that wretched place that I had been Haroun, or had had a Seraglio, for, I knew that if I mentioned my reverses, I should be so worried, that I should have to drown myself in the muddy pond near the playground, which looked like the beer.

Culprit – *Offender*
Flutter – *Flap*
Oblige – *To bind morally or legally*
Pursue – *Follow*

Ah me, ah me! No other ghost has haunted the boy's room, my friends, since I have occupied it, than the ghost of my own childhood, the ghost of my own innocence, the ghost of my own airy belief. Manyatimes have I pursued the phantom, never with this man's **stride** of mine to come up with it, never with these man's hands of mine to touch it, never more to this man's heart of mine to hold it in its purity. And here you see me working out, as cheerfully and thankfully as I may, my doom of shaving in the glass a constant change of customers, and of lying down and rising up with the skeleton allotted to me for my **mortal companion**.

Food For Thought

'The Haunted House' is basically a story which was started and ended by Charles Dickens but has inputs from a number of authors, such as Wilkie Collins, Hesba Stretton, Elizabeth Gaskell, etc. What do you feel after reading the whole stroy? Does the story contain only terror or mystery or is there an element of humours also in it? Explain your views about the story with relevant reasons.

Stride – *Pace*
Mortal – *Human*
Companion –
Friend

An Understanding

Q. 1. What happens when the narrator sees a deserted house from his railway carriage?

Ans. _____

_____

_____

_____

_____

Q. 2. What does the narrator decide to do seeing the deserted house? What was so different about that particular house?

Ans. _____

_____

_____

_____

_____

Q. 3. What did the narrator do with the servants of the haunted house? What happened during the Cristmas Eve?

Ans. _____

_____

_____

_____

_____

Q. 4. Was the house really haunted? What happened during the Twelfth Night?

Ans. _____

_____

_____

_____

_____

Ambrose Gwinnett Bierce

Born on June 24, 1842

Died on sometime after December 26, 1913

Notable Works: *The Fiend's Delight, The Devil's Dictionary, The Cynic's Word Book, Collected Works* and a number of ghost stories and realistic and short war stories, such as: 'An Occurrence at Owl Creek Bridge', 'The Boarded Window', 'Killed at Resaca', and 'Chickamauga'. He also published several volumes of poetry like the *Fantastic Fables.* He published a column called 'Prattle' and became one of the first regular columnists and editorialists to be employed on William Randolph Hearst's newspaper, the *San Francisco Examiner*,

Honours: At least three films have been made of Bierce's story, 'An Occurrence at Owl Creek Bridge'. A silent film version, *The Bridge*, was made in 1929. A French version called *La Rivière du Hibou*, directed by Robert Enrico, was released in 1962 and another version, directed by Brian James Egen, was released in 2005.

Early life

Ambrose Gwinnett Bierce, an American editorialist, journalist, short story writer, fabulist and satirist was born on June 24, 1842. Bierce was born at Horse Cave Creek in Meigs County, Ohio to Marcus Aurelius Bierce and Laura Sherwood Bierce. His parents were a poor, but literary couple who instilled in him a deep love for books and writing. The boy grew up in Kosciusko County, Indiana, attending high school at the county seat, Warsaw.

Military Career

At the outset of the American Civil War, Bierce enlisted in the Union Army's 9th Indiana Infantry Regiment. He participated in the Operations in Western Virginia campaign (1861), and was present at the "first battle" at Philippi. He received newspaper attention for his daring rescue, under fire, of a gravely wounded comrade at the Battle of Rich Mountain. In February 1862, he was commissioned First Lieutenant, and served on the staff of General William Babcock Hazen as a topographical engineer, making maps of likely battlefields.

Bierce fought at the Battle of Shiloh (April 1862), a terrifying experience that became a source for several later short stories and the memoir, 'What I Saw of Shiloh'. Bierce received the rank of *brevet major* before resigning from the Army.

Journalistic Career

Bierce remained in San Francisco for many years, eventually becoming famous as a contributor and/or editor for a number of local newspapers and periodicals, including

The San Francisco News Letter, *The Argonaut*, the *Overland Monthly*, *The Californian* and *The Wasp*. A selection of his crime reporting from *The San Francisco News Letter* was included in The Library of America anthology *True Crime*.

Literary Works and Achievements

Bierce was considered **a master of pure English** by his contemporaries, and virtually everything that came from his pen was notable for its judicious wording and economy of style. He wrote in a variety of literary genres.

Bierce lived and wrote in England from 1872 to 1875, contributing to *Fun* magazine. His first book, *The Fiend's Delight*, a compilation of his articles, was published in London in 1873 by John Camden Hotten under the pseudonym, "Dod Grile. Returning to the United States, he again took up residence in San Francisco. From 1879 to 1880, he travelled to Rockerville and Deadwood in the Dakota Territory, to try his hand as local manager for a New York mining company, but when the company failed he, returned to San Francisco and resumed his career in journalism.

In 1887, he published a column called 'Prattle' and became one of the first regular columnists and editorialists to be employed on William Randolph Hearst's newspaper, the *San Francisco Examiner*, eventually becoming one of the most prominent and influential among the writers and journalists of the West Coast. He remained associated with Hearst Newspapers until 1906.

His short stories are held among the best of the 19th century, providing a popular following based on his roots. He wrote realistically of the terrible things he had seen in the war in such stories as "An Occurrence at Owl Creek Bridge", "The Boarded Window", "Killed at Resaca", and "Chickamauga". In addition to his ghost and war stories, he also published several volumes of poetry. His *Fantastic Fables* anticipated the ironic style of grotesquerie that became a more common genre in the 20th century.

One of Bierce's most famous works is his much-quoted book, *The Devil's Dictionary*, originally an occasional newspaper item which was first published in book form in 1906 as *The Cynic's Word Book*. It consists of satirical definitions of English words and political double-talk.Bierce's twelve-volume, *Collected Works* were published in 1909, the seventh volume of which consists solely of *The Devil's Dictionary*, the title Bierce himself preferred to *The Cynic's Word Book*.

Writing Style

Despite his reputation as a searing critic, Bierce was known to encourage younger writers, including poet George Sterling and fiction writer W. C. Morrow. Bierce had a distinctive style of writing, especially in his stories. His style often embraces **an abrupt beginning, dark imagery, vague references to time, limited descriptions, impossible events** and the **theme of war.**

Disappearance

In October 1913, the 71-year old Bierce, departed to Washington, D.C., for a tour of his old Civil War battlefields. By December, he had proceeded through Louisiana and Texas, crossing by way of El Paso into Mexico, which was in the throes of revolution. Bierce travelled to Mexico to gain a first-hand experience of the Mexican Revolution. While traveling with the rebel troops, he disappeared without a trace.

Trivia

Ambrose Gwinnett Bierce was the tenth of the thirteen children that his parents had, and his father gave all of them names beginning with the letter, "A".

One Of Twins

– Ambrose Bierce

A letter found among the papers of the late Mortimer Barr

YOu ask me if in my experience as one of a pair of twins I ever observed anything unaccountable by the natural laws with which we have acquaintance. As to that you shall judge; perhaps we have not all acquaintance with the same natural laws. You may know some that I do not, and what is to me unaccountable may be very clear to you.

You knew my brother John -- that is, you knew him when you knew that I was not present; but neither you nor, I believe, any human being could distinguish between him and me if we chose to seem alike. Our parents could not; ours is the only instance of which I have any knowledge of so close resemblance as that. I speak of my brother John, but I am not at all sure that his name was not Henry and mine John. We were regularly christened, but afterward, in the very act of tattooing us with small distinguishing marks, the operator lost his reckoning; and although I bear upon my forearm a small 'H' and he bore a 'J,' it is by no means certain that the letters ought not to have been transposed. During our boyhood our parents tried to distinguish us more obviously by our clothing and other simple devices, but we would so frequently exchange suits and otherwise circumvent the enemy that they abandoned all such ineffectual attempts, and during all the years that we lived together at home everybody recognized the difficulty of the situation and made the best of it by calling us both 'Jehnry.' I have often wondered at my father's forbearance in not branding us conspicuously upon our unworthy brows, but as we were tolerably good boys and used our power of embarrassment and annoyance with commendable moderation, we escaped the iron. My father was, in fact, a singularly good-natured man, and I think quietly enjoyed nature's practical joke.

Soon after we had come to California, and settled at San Jose (where the only good fortune that awaited us was our meeting with so kind a friend as you), the family, as you know, was broken up by the death of both my parents in the same week. My father died insolvent, and the homestead was

Acquaintance – *Associate,*
Desert - *Forsake utterly*
Distinguish – *Differentiate*
Commendable – *Praiseworthy*
Insolvent *– Bankrupt, Pneniless*

sacrificed to pay his debts. My sisters returned to relatives in the East, but owing to your kindness John and I, then twenty-two years of age, obtained employment in San Francisco, in different quarters of the town. Circumstances did not permit us to live together, and we saw each other infrequently, sometimes not oftener than once a week. As we had few acquaintances in common, the fact of our extraordinary likeness was little known. I come now to the matter of your inquiry.

One day soon after we had come to this city I was walking down Market Street late in the afternoon, when I was accosted by a well-dressed man of middle age, who after greeting me cordially said, 'Stevens, I know, of course, that you do not go out much, but I have told my wife about you, and she would be glad to see you at the house. I have a notion, too, that my girls are worth knowing. Suppose you come out tomorrow at six and dine with us, en famille; and then if the ladies can't amuse you afterwards I'll stand in with a few games of billiards.'

This was said with so bright a smile and so engaging a manner that I had not the heart to refuse, and although I had never seen the man in my life I promptly replied, 'You are very good, sir, and it will give me great pleasure to accept the invitation. Please present my compliments to Mrs. Margovan and ask her to expect me.'

With a shake of the hand and a pleasant parting word, the man passed on. That he had mistaken me for my brother was plain enough. That was an error to which I was accustomed and which it was not my habit to rectify unless the matter seemed important. But how had I known that this man's name was Margovan? It certainly is not a name that one would apply to a man at random, with a probability that it would be right. In point of fact, the name was as strange to me as the man.

The next morning I hastened to where my brother was employed and met him coming out of the office with a number of bills that he was to collect. I told him how I had 'committed' him and added that if he didn't care to keep the engagement I should be delighted to continue the impersonation. 'That's queer,' he said thoughtfully. 'Margovan is the only man in the office here whom I know well and like. When he came in this morning and we had passed the usual greetings some singular impulse prompted me to say, "Oh, I beg your pardon,

Debt – *Dues*
Permit – *License, To allow*
Pleasure – *Desire*
Rectify – *Correct*

Greatest   Science   Fiction   Stories

Mr. Margovan, but I neglected to ask your address." I got the address, but what under the sun I was to do with it, I did not know until now. It's good of you to offer to take the consequence of your impudence, but I'll eat that dinner myself, if you please.'

He ate a number of dinners at the same place -- more than were good for him, I may add without disparaging their quality; for he fell in love with Miss Margovan, proposed marriage to her and was heartlessly accepted.

Several weeks after I had been informed of the engagement, but before it had been convenient for me to make the acquaintance of the young woman and her family, I met one day on Kearney Street a handsome but somewhat dissipated-looking man. Something prompted me to follow and watch, which I did without any scruple whatever. He turned up Geary Street and followed it until he came to Union Square. There he looked at his watch, then entered the square. He loitered about the paths for some time, evidently waiting for some one. Presently he was joined by a fashionably dressed and beautiful young woman and the two walked away up Stockton Street, with me following. I now felt the necessity of extreme caution, for although the girl was a stranger, it seemed to me that she would recognize me at a glance. They made several turns from one street to another and finally, after both had taken a hasty look all about -- which I narrowly evaded by stepping into a doorway -- they entered a house of which I do not care to state the location. Its location was better than its character.

I protest that my action in playing the spy upon these two strangers was without assignable motive. It was one of which I might or might not be ashamed, according to my estimate of the character of the person finding it out. As an essential part of a narrative educed by your question it is related here without hesitancy or shame.

A week later John took me to the house of his prospective father-in-law, and in Miss Margovan, as you have already surmised, but to my profound astonishment, I recognized the heroine of that discreditable adventure. A gloriously beautiful heroine of a discreditable adventure I must in justice admit that she was; but that fact has only this importance: her beauty was such a surprise to me that it cast a doubt upon her identity with the young woman I had seen before; how could the

Impudence –
Boldness
Scruple – *Restraint*
Misgiving
Loiter – *Wander*
Caution – *Carefulness*
Surmise – *Imagine,*
Suspect, Guess

marvellous fascination of her face have failed to strike me at that time? But no -- there was no possibility of error; the difference was due to costume, light and general surroundings.

John and I passed the evening at the house, enduring, with the fortitude of long experience, such delicate enough banter as our likeness naturally suggested. When the young lady and I were left alone for a few minutes I looked her squarely in the face and said with sudden gravity,

'You, too, Miss Margovan, have a double. I saw her last Tuesday afternoon in Union Square.'

She trained her great grey eyes upon me for a moment, but her glance was a trifle less steady than my own and she withdrew it, fixing it on the tip of her shoe.

'Was she very like me?' she asked, with an indifference which I thought a little overdone.

'So like,' said I, 'that I greatly admired her, and being unwilling to lose sight of her I confess that I followed her until -- Miss Margovan, are you sure that you understand?'

She was now pale, but entirely calm. She again raised her eyes to mine, with a look that did not falter.

'What do you wish me to do?' she asked. 'You need not fear to name your terms. I accept them.'

It was plain, even in the brief time given me for reflection, that in dealing with this girl ordinary methods would not do, and ordinary exactions were needless.

'Miss Margovan,' I said, doubtless with something of the compassion in my voice that I had in my heart, 'it is impossible not to think you the victim of some horrible compulsion. Rather than impose new embarrassments upon you I would prefer to aid you to regain your freedom.'

She shook her head, sadly and hopelessly, and I continued, with agitation,

'Your beauty unnerves me. I am disarmed by your frankness and your distress. If you are free to act upon conscience you will, I believe, do what you conceive to be best; if you are not -- well, Heaven help us all! You have nothing to fear from me but such opposition to this marriage as I can try to justify on -- on other grounds.'

These were not my exact words, but that was the sense of them, as nearly as my sudden and conflicting emotions

Fortitude – *Strength*
Banter – *Teasing*
Aid – *Help*
Distress – *Pain*
Justify – *Defend*

permitted me to express it. I rose and left her without another look at her, met the others as they re-entered the room and said, as calmly as I could, 'I have been bidding Miss Margovan good evening; it is later than I thought.'

John decided to go with me. In the street he asked if I had observed anything singular in Julia's manner.

'I thought her ill,' I replied; 'that is why I left.' Nothing more was said.

The next evening I came late to my lodgings. The events of the previous evening had made me nervous and ill; I had tried to cure myself and attain to clear thinking by walking in the open air, but I was oppressed with a horrible presentiment of evil -- a presentiment which I could not formulate. It was a chill, foggy night; my clothing and hair were damp and I shook with cold. In my dressing gown and slippers before a blazing grate of coals I was even more uncomfortable. I no longer shivered but shuddered -- there is a difference. The dread of some impending calamity was so strong and dispiriting that I tried to drive it away by inviting a real sorrow -- tried to dispel the conception of a terrible future by substituting the memory of a painful past. I recalled the death of my parents and endeavoured to fix my mind upon the last sad scenes at their bedsides and their graves. It all seemed vague and unreal, as having occurred ages ago and to another person. Suddenly, striking through my thought and parting it as a tense cord is parted by the stroke of steel -- I can think of no other comparison -- I heard a sharp cry as of one in mortal agony! The voice was that of my brother and seemed to come from the street outside my window. I sprang to the window and threw it open. A street lamp directly opposite threw a wan and ghastly light upon the wet pavement and the fronts of the houses. A single policeman, with upturned collar, was leaning against a gatepost, quietly smoking a cigar. No one else was in sight. I closed the window and pulled down the shade, seated myself before the fire and tried to fix my mind upon my surroundings. By way of assisting, by performance of some familiar act, I looked at my watch; it marked half-past eleven. Again I heard that awful cry! It seemed in the room -- at my side. I was frightened and for some moments had not the power to move. A few minutes later -- I have no recollection of the intermediate

Oppress – *Supress, Dominate*

Dread – *Fear*

Dispel – *Dismiss, Alleviate*

Wan – *Pale*

Intermediate – *Middle*

time -- I found myself hurrying along an unfamiliar street as fast as I could walk. I did not know where I was, nor whither I was going, but presently sprang up the steps of a house before which were two or three carriages and in which were moving lights and a subdued confusion of voices. It was the house of Mr. Margovan.

You know, good friend, what had occurred there. In one chamber lay Julia Margovan, hours dead by poison; in another John Stevens, bleeding from a pistol wound in the chest, inflicted by his own hand. As I burst into the room; pushed aside the physicians and laid my hand upon his forehead, he unclosed his eyes, stared blankly, closed them slowly and died without a sign.

I knew no more until six weeks afterwards, when I had been nursed back to life by your own saintly wife in your own beautiful home. All of that you know, but what you do not know is this -- which, however, has no bearing upon the subject of your psychological researches -- at least not upon that branch of them in which, with a delicacy and consideration all your own, you have asked for less assistance than I think I have given you,

One moonlight night several years afterwards I was passing through Union Square. The hour was late and the square deserted. Certain memories of the past naturally came into my mind as I came to the spot where I had once witnessed that fateful assignation, and with that unaccountable perversity which prompts us to dwell upon thoughts of the most painful character I seated myself upon one of the benches to indulge them. A man entered the square and came along the walk towards me. His hands were clasped behind him, his head was bowed; he seemed to observe nothing. As he approached the shadow in which I sat I recognized him as the man whom I had seen meet Julia Margovan years before at that spot. But he was terribly altered -- grey, worn, and haggard. Dissipation and vice were in evidence in every look; illness was no less apparent. His clothing was in disorder, his hair fell across his forehead in a derangement which was at once uncanny, and picturesque. He looked fitter for restraint than liberty -- the restraint of a hospital.

Subdue – *Pacify*
Inflict – *Impose*
Dwell – *Reside*
Indulge – *Pamper*
Vice – *Evil*
Uncanny – *Weird*

With no defined purpose I rose and confronted him. He raised his head and looked me full in the face. I have no words to describe the ghastly change that came over his own; it was a look of unspeakable terror -- he thought himself eye to eye with a ghost. But he was a courageous man. 'Damn you, John Stevens!' he cried, and lifting his trembling arm he dashed his fist feebly at my face and fell headlong upon the gravel as I walked away.

Somebody found him there, stone-dead. Nothing more is known of him, not even his name. To know of a man that he is dead should be enough.

Food For Thought

Who do you think killed Henry's twin brother, John Stevens and why? How did Julia Margovan die? What happened several years later in a moon-lit night? What message do you get from the story?

Feeble – *Weak*

An Understanding

Q. 1. What were the names of the twins? What did their parents do in their boyhood to distinguish them? What else did they have to identify each one from the other?

Ans. _____

_____

_____

_____

_____

Q. 2. What were the circumstances under which John and Henry came to California? Where did they settle and what did they do there?

Ans. _____

_____

_____

_____

_____

Q. 3. Who was Mr. Margovan? Where did Henry meet him and why did Mr. Margovan address him as Stevens, inviting him for dinner?

Ans. _____

_____

_____

_____

_____

Q. 4. Who actually, went to the dinner at Mr. Margovan's house? What happened there? Who was supposed to marry Miss Margovan and what happened at the end?

Ans. _____

_____

_____

_____

_____

Moxon's Master

—Ambrose Bierce

'Are you serious? -- do you really believe that a machine thinks?'

I got no immediate reply; Moxon was apparently intent upon the coals in the grate, touching them deftly here and there with the fire-poker till they signified a sense of his attention by a brighter glow. For several weeks I had been observing in him a growing habit of delay in answering even the most trivial of commonplace questions. His air, however, was that of preoccupation rather than deliberation: one might have said that he had 'something on his mind.'

Presently he said,

'What is a "machine"? The word has been variously defined. Here is one definition from a popular dictionary: "Any instrument or organization by which power is applied and made effective, or a desired effect produced." Well, then, is not a man a machine? And you will admit that he thinks -- or thinks he thinks.'

'If you do not wish to answer my question,' said, rather testily, 'why not say so? -- all that you say is mere evasion. You know well enough that when I say "machine" I do not mean a man, but something that man has made and controls.'

'When it does not control him,' he said, rising abruptly and looking out of a window, whence nothing was visible in the blackness of a stormy night. A moment later he turned about and with a smile said, 'I beg your pardon; I had no thought of evasion. I considered the dictionary man's unconscious testimony suggestive and worth something in the discussion. I can give your question a direct answer easily enough, I do believe that a machine thinks about the work that it is doing.'

That was direct enough, certainly. It was not altogether pleasing, for it tended to confirm a sad suspicion that Moxon's devotion to study and work in his machine-shop had not been good for him. I knew, for one thing, that he suffered from insomnia, and that is no light affliction. Had it affected his mind? His reply to my question seemed to me then evidence that it had; perhaps I should think differently about it now. I was younger then, and among the blessings that are not

Apparently - *Easily understable*
Fire-poker - *A metal instrument used to handle fire or charcoal*
Preoccupation - *Being very busy*
Abruptly - *Suddenly*
Machine-Shop - *A workshop in which metals are cut and shaped*

denied to youth is ignorance. Incited by that great stimulant to controversy, I said,

'And what, pray, does it think with -- in the absence of a brain?'

The reply, coming with less than his customary delay, took his favourite form of counter-interrogation, 'With what does a plant think -- in the absence of a brain?'

'Ah, plants also belong to the philosopher class! I should be pleased to know some of their conclusions; you may omit the premises.'

'Perhaps,' he replied, apparently unaffected by my foolish irony, 'you may be able to infer their convictions from their acts. I will spare you the familiar examples of the sensitive mimosa, the several insectivorous flowers and those whose stamens bend down and shake their pollen upon the entering bee in order that it may fertilize their distant mates. But observe this. In an open spot in my garden I planted a climbing vine. When it was barely above the surface I set a stake into the soil a yard away.

The vine at once made for it, but as it was about to reach it after several days I removed it a few feet. The vine at once altered its course, making an acute angle, and again made for the stake. This manoeuvre was repeated several times, but finally, as if discouraged, the vine abandoned the pursuit and ignoring further attempts to divert it, travelled to a small tree, farther away, which it climbed.

'Roots of the eucalyptus will prolong themselves incredibly in search of moisture. A well-known horticulturist relates that one entered an old drain pipe and followed it until it came to a break, where a section of the pipe had been removed to make way for a stone wall that had been built across its course. The root left the drain and followed the wall until it found an opening where a stone had fallen out. It crept through and following the other side of the wall back to the drain, entered the unexplored part and resumed its journey.'

'And all this?'

'Can you miss the significance of it? It shows the consciousness of plants. It proves that they think.' 'Even if it did -- what then? We were speaking, not of plants, but of machines. They may be composed partly of wood -- wood that

Vine - *A grape plant*
Ignorance - *Lack of knowledge*
Horticulturist - *A person studying the science of cultivating plants*
Manoeuvre - *Deceptive plan/action*

has no longer vitality -- or wholly of metal. Is thought also an attribute of the mineral kingdom?'

'How else do you explain the phenomena, for example, of crystallization?'

'I do not explain them.'

'Because you cannot without affirming what you wish to deny, namely, intelligent co-operation, among the constituent elements of the crystals. When soldiers form lines, or hollow squares, you call it reason. When wild geese in flight take the form of a letter V you say instinct. When the homogeneous atoms of a mineral, moving freely in solution, arrange themselves into shapes mathematically perfect, or particles of frozen moisture into the symmetrical and beautiful forms of snowflakes, you have nothing to say. You have not even invented a name to conceal your heroic unreason.'

Moxon was speaking with unusual animation and earnestness. As he paused I heard in an adjoining room known to me as his 'machine-shop,' which no one but he himself was permitted to enter, a singular thumping sound, as of someone pounding upon a table with an open hand. Moxon heard it at the same moment and, visibly agitated, rose and hurriedly passed into the room whence it came.

I thought it odd that anyone else should be in there, and my interest in my friend -- with doubtless a touch of unwarrantable curiosity -- led me to listen intently, though, I am happy to say, not at the keyhole. There were confused sounds, as of a struggle or scuffle; the floor shook.

I distinctly heard hard breathing and a hoarse whisper which said 'Damn you!' Then all was silent, and presently Moxon reappeared and said, with a rather sorry smile.

'Pardon me for leaving you so abruptly. I have a machine in there that lost its temper and cut up rough.'

Fixing my eyes steadily upon his left cheek, which was traversed by four parallel excoriations showing blood, I said,

Crystallisation - *To form/cause to form crystals*
Symmetrical - *Well-proportioned*
Vitality - *Vigour*
Attribute - *Assign, Impute to*

'How would it do to trim its nails?' I could have spared myself the jest; he gave it no attention, but seated himself in the chair that he had left and resumed the interrupted monologue as if nothing had occurred,

'Doubtless you do not hold with those (I need not name them to a man of your reading) who have taught that every matter is sentient, that every atom is a living, feeling, conscious being. I do. There is no such thing as dead, inert matter.

It is all alive; all instinct with force, actual and potential; all sensitive to the same forces in its environment and susceptible to the contagion of higher and subtler ones residing in such superior organisms as it may be brought into relation with, as those of man when he is fashioning it into an instrument of his will.

It absorbs something of his intelligence and purpose -- more of them in proportion to the complexity of the resulting machine and that of its work.

'Do you happen to recall Herbert Spencer's definition of "Life"? I read it thirty years ago. He may have altered it afterward, for anything I know, but in all that time I have been unable to think of a single word that could profitably be changed or added or removed. It seems to me not only the best definition, but the only possible one.

'"Life," he says, "is a definite combination of heterogeneous changes, both simultaneous and successive, in correspondence with external co-existences and sequences."'

'That defines the phenomenon,' I said, 'but gives no hint of its cause.'

'That,' he replied, 'is all that any definition can do. As Mill points out, we know nothing of cause except as an antecedent -- nothing of effect except as a consequent. Of certain phenomena, one never occurs without another, which is dissimilar: the first in point of time we call cause, the second, effect. One who had many times seen a rabbit pursued by a dog, and had never seen rabbits and dogs otherwise, would think the rabbit the cause of the dog.

'But I fear,' he added, laughing naturally enough, 'that my rabbit is leading me a long way from the track of my legitimate quarry. I'm indulging in the pleasure of the chase for its own sake.

What I want you to observe is that in Herbert Spencer's definition of "life" the activity of a machine is included -- there is nothing in the definition that is not applicable to it. According to this sharpest of observers and deepest of thinkers, if a man during his period of activity is alive, so is a machine when in operation. As an inventor and constructor of machines I know that to be true.'

Contagion - *A contagious disease*
Antecedent - *Preceding*
Indulging - *Allowing*
Sentient - *Having senses, Conscious*

Moxon was silent for a long time, gazing absently into the fire. It was growing late and I thought it time to be going, but somehow I did not like the notion of leaving him in that isolated house, all alone except for the presence of some person of whose nature my conjectures could go no further than that it was unfriendly, perhaps malign. Leaning towards him and looking earnestly into his eyes while making a motion with my hand through the door of his workshop, I said,

'Moxon, whom have you in there?'

Somewhat to my surprise he laughed lightly and answered without hesitation,

'Nobody; the incident that you have in mind was caused by my folly in leaving a machine in action with nothing to act upon, while I undertook the interminable task of enlightening your understanding. Do you happen to know that consciousness is the creature of rhythm?'

'O bother them both!' I replied, rising and laying hold of my overcoat. 'I'm going to wish you good night; and I'll add the hope that the machine which you inadvertently left in action will have her gloves on the next time you think it needful to stop her.'

Without waiting to observe the effect of my shot I left the house.

Rain was falling, and the darkness was intense. In the sky beyond the crest of a hill towards which I groped my way along precarious plank sidewalks and across miry, unpaved streets I could see the faint glow of the city's lights, but behind me nothing was visible but a single window of Moxon's house. It glowed with what seemed to me a mysterious and fateful meaning.

I knew it was an uncurtained aperture in my friend's 'machine-shop,' and I had little doubt that he had resumed the studies interrupted by his duties as my instructor in mechanical consciousness and the fatherhood of rhythm. Odd, and in some degree humorous, as his convictions seemed to me at that time,

Hesitation - *Indecision*

Folly - *Mistake*

Malign - *Defame*

Conviction - *Fixed/ Firm beliefs*

Divest - *To sell off, Strip, Deprive*

I could not wholly divest myself of the feeling that they had some tragic relation to his life and character -- perhaps to his destiny -- although I no longer entertained the notion that they were the vagaries of a disordered mind. Whatever might be thought of his views, his exposition of them was too logical for that. Over and over, his last words came back to me, 'consciousness is the creature of rhythm.' Bald and terse as the statement was, I now found it infinitely alluring.

At each recurrence it broadened in meaning and deepened in suggestion. Why, here (I thought) is something upon which to found a philosophy. If consciousness is the product of rhythm all things are conscious, for all have motion, and all motion is rhythmic. I wondered if Moxon knew the significance and breadth of his thought -- the scope of this momentous generalization; or had he arrived at his philosophic faith by the tortuous and uncertain road of observation?

That faith was then new to me, and all Moxon's expounding had failed to make me a convert; but now it seemed as if a great light shone about me, like that which fell upon Saul of Tarsus; and out there in the storm and darkness and solitude I experienced what Lewes calls 'The endless variety and excitement of philosophic thought.' I exulted in a new sense of knowledge, a new pride of reason. My feet seemed hardly to touch the earth; it was as if I were uplifted and borne through the air by invisible wings.

Yielding to an impulse to seek further light from him whom I now recognized as my master and guide, I had unconsciously turned about, and almost before I was aware of having done so found myself again at Moxon's door. I was drenched with rain, but felt no discomfort. Unable in my excitement to find the doorbell I instinctively tried the knob. It turned and, entering, I mounted the stairs to the room that I had so recently left.

All was dark and silent; Moxon, as I had supposed, was in the adjoining room -- the 'machine-shop.' Groping along the wall until found the communicating door I knocked loudly several times, but got no response, which I attributed to the uproar outside, for the wind was blowing a gale and dashing the rain against the thin walls in sheets.

The drumming upon the shingle roof spanning the unceiled room was loud and incessant. I had never been invited into the machine-shop, indeed, had been denied admittance, as had all others, with one exception, a skilled metal worker, of whom no one knew anything except that his name was Haley and his habit silence. But in my spiritual exaltation, discretion

Destiny - *Fate*
Alluring - *Tempting*
Instinctively - *Spontaneously*
Groping - *To feel about with the hands*
Gale - *A very strong wind*

and civility were alike forgotten, and I opened the door. What I saw took all philosophical speculation out of me in short order.

Moxon sat facing me at the farther side of a small table upon which a single candle made all the light that was in the room. Opposite him, his back toward me, sat another person. On the table between the two was a chess-board; the men were playing. I knew little of chess, but as only a few pieces were on the board it was obvious that the game was near its close.

Moxon was intensely interested -- not so much, it seemed to me, in the game as in his antagonist, upon whom he had fixed so intent a look that, standing though I did directly in the line of his vision, I was altogether unobserved. His face was ghastly white, and his eyes glittered like diamonds. Of his antagonist I had only a back view, but that was sufficient; I should not have cared to see his face.

He was apparently not more than five feet in height, with proportions suggesting those of a gorilla -- a tremendous breadth of shoulders, thick, short neck and broad, squat head, which had a tangled growth of black hair and was topped with a crimson fez. A tunic of the same colour, belted tightly to the waist, reached the seat -- apparently a box -- upon which he sat; his legs and feet were not seen. His left forearm appeared to rest in his lap; he moved his pieces with his right hand, which seemed disproportionately long.

I had shrunk back and now stood a little to one side of the doorway and in shadow. If Moxon had looked farther than the face of his opponent he could have observed nothing now, except that the door was open. Something forbade me either to enter or to retire, a feeling -- I know not how it came -- that I was in the presence of an imminent tragedy and might serve my friend by remaining. With a scarcely conscious rebellion against the indelicacy of the act, I remained.

The play was rapid. Moxon hardly glanced at the board before making his moves, and to my unskilled eye seemed to move the piece most convenient to his hand, his motions in doing so being quick, nervous, and lacking in precision. The response of his antagonist, while equally prompt in inception, was made with a slow, uniform, mechanical and, I thought, somewhat theatrical movement of the arm that was a sore trial to my patience. There was something unearthly about it all, and I caught myself shuddering. But I

Exception - *Special/ Different from other*

Speculation - *Contemplation, Consideration*

Disproportion - *Unequal, Unsymmetical*

Imminent- *Likely to occur, Impending*

Antagonist - *Opponent, the opposite of a Hero in a story/ drama*

was wet and cold. Two or three times after moving a piece, the stranger slightly inclined his head, and each time I observed that Moxon shifted his king.

All at once the thought came to me that the man was dumb. And then that he was a machine -- an automaton chess player! Then I remembered that Moxon had once spoken to me of having invented such a piece of mechanism, though

I did not understand that it had actually been constructed. Was all his talk about the consciousness and intelligence of machines merely a prelude to eventual exhibition of this device -- only a trick to intensify the effect of its mechanical action upon me in my ignorance of its secret?

A fine end, this, of all my intellectual transports -- my 'endless variety and excitement of philosophic thought'! I was about to retire in disgust when something occurred to hold my curiosity.

I observed a shrug of the thing's great shoulders, as if it were irritated, and so natural was this -- so entirely human -- that in my new view of the matter it startled me. Nor was that all, for a moment later it struck the table sharply with its clenched hand. At that gesture Moxon seemed even more startled than I. He pushed his chair a little backward, as in alarm.

Presently Moxon, whose play it was, raised his hand high above the board, pounced upon one of his pieces like a sparrow-hawk and with the exclamation 'check-mate!' rose quickly to his feet and stepped behind his chair. The automaton sat motionless.

The wind had now gone down, but I heard, at lessening intervals and progressively louder, the rumble and roll of thunder. In the pauses between I now became conscious of a low humming or buzzing which, like the thunder, grew momentarily louder and more distinct.

It seemed to come from the body of the automaton, and was unmistakably a whirring of wheels. It gave me the impression of a disordered mechanism which had escaped the repressive and regulating action of some controlling part -- an effect such as might be expected if a pawl should be jostled from the teeth of a ratchetwheel. But before I had time for

Shuddering -
Trembling
Intensify -
Strengthen, Deepen
Prelude -
Introduction
Whirring -
Revolving/Moving with a buzzing sound

much conjecture as to its nature, my attention was taken by the strange motions of the automaton itself.

A slight but continuous convulsion appeared to have possession of it. In body and head it shook like a man with palsy or an ague chill, and the motion augmented every moment until the entire figure was in violent agitation. Suddenly it sprang to its feet and with a movement almost too quick for the eye to follow shot forward across table and chair, with both arms thrust forth to their full length -- the posture and lunge of a diver. Moxon tried to throw himself backward out of reach, but he was too late.

I saw the horrible thing's hand close upon his throat, his own clutch its wrists. Then the table was overturned, and candle thrown to the floor and extinguished, and all was black dark. But the noise of the struggle was dreadfully distinct, and most terrible of all were the raucous, squawking sounds made by the strangled man's efforts to breathe.

Guided by the infernal hubbub, I sprang to the rescue of my friend, but had hardly taken a stride in the darkness when the whole room blazed with a blinding white light that burned into my brain and heart and memory a vivid picture of the combatants on the floor, Moxon underneath, his throat still in the clutch of those iron hands, his head forced backward, his eyes protruding, his mouth wide open and his tongue thrust out; and -- horrible contrast! -- upon the painted face of his assassin an expression of tranquil and profound thought, as in the solution of a problem in chess! This I observed, then all was blackness and silence.

Three days later I recovered consciousness in a hospital. As the memory of that tragic night slowly evolved in my ailing brain I recognized in my attendant Moxon's confidential workman, Haley. Responding to a look he approached, smiling.

'Tell me about it,' I managed to say, faintly -- 'all about it.'

'Certainly,' he said; 'you were carried unconscious from a burning house -- Moxon's. Nobody knows how you came to be there. You may have to do a little explaining. The origin of the fire is a bit mysterious, too. My own notion is that the house was struck by lightning.'

'And Moxon?'

'Buried yesterday -- what was left of him.'

Infernal - *Devilish*
Hubbub - *A loud, confused noise*
Conjecture - *An opinion/theory*
Augmental - *To make larger in size, strength, number, etc.*

Apparently this reticent person could unfold himself on occasion. When imparting shocking intelligence to the sick, he was affable enough. After some moments of the keenest mental suffering I ventured to ask another question,

'Who rescued me?'

'Well, if that interests you -- I did.'

'Thank you, Mr. Haley, and may God bless you for it. Did you rescue, also, that charming product of your skill, the automaton chess player that murdered its inventor?'

The man was silent a long time, looking away from me. Presently he turned and gravely said,

'Do you know that?'

'I do,' I replied; 'I saw it done.'

That was many years ago. If asked today I should answer less confidently.

Food For Thought

Why do you think the narrator in the story question at the end that whatever he saw was real or not? Moxon wins the game of chess but the automaton kills him after losing the game in anger. What does this signify and indicate?

Affable - *Cordial*
Automation - *Robot*
Apparently - *Evidently*
Ventured - *To undertake, Embark*

An Understanding

Q. 1. What is the main theme of the story and what is an automaton? Which type of automaton has been used in this story?

Ans. _____

_____

_____

_____

_____

Q. 2. Who is Moxon and what is the narrator conversing with him about?

Ans. _____

_____

_____

_____

Q. 3. What was Moxon doing with his automaton when the narrator returns to Moxon's house?

Ans. _____

_____

_____

_____

Q. 4. How was Moxon killed? Who killed him and why?

Ans. _____

_____

_____

_____

www.ingramcontent.com/pod-product-compliance
Lightning Source LLC
Chambersburg PA
CBHW061304220326

41599CB00026B/4729